what
your
cat
wants

what your cat wants

FRANCESCA RICCOMINI

hamlyn

An Hachette UK Company
www.hachette.co.uk

First published in Great Britain
in 2012 by
Hamlyn, a division of Octopus
Publishing Group Ltd
Endeavour House
189 Shaftesbury Avenue
London
WC2H 8JY
www.octopusbooks.co.uk

ISBN 978-0-600-62462-2

A CIP catalogue record for this
book is available from the British
Library

Printed and bound in China

10 9 8 7 6 5 4 3 2 1

Acknowledgements

Executive editor: Trevor Davies
Designed and produced by SP Creative Design
Editor: Heather Thomas
Designer: Rolando Ugolini
Special photography: Rolando Ugolini

Picture credits
The publishers would like to thank the following
individuals for their assistance in the photography: Marcia
and John Castle and Susan Mortimer.

Octopus Publishing Group Limited/Rolando Ugolini 7, 10
bottom, 11 bottom, 13 top, 14 top, 15 top, 17, 19 top,
20 left, 21 top, 23, 24, 28 top, 32 top, 35, 36, 37 bottom,
44, 45, 47 right, 52, 57, 58, 59, 61, 67 left, 69 bottom, 70
bottom, 71, 73 bottom, 74, 76, 77, 78, 79 right, 81, 82,
84, 86 bottom, 90, 91 right, 93

Octopus Publishing Group Limited 41; Octopus Publishing
Group Limited/Jane Burton 61 bottom, 80, 92 top, 92
bottom; /Keith Colin 92; /John Daniels 20 right; /Steve
Gorton 18; /Ray Moller 15 bottom, 55 right; /Russell Sadur
11 right

Rolando Ugolini 9, 10 top, 14 bottom, 16 bottom, 19
bottom, 22, 25 top, 26, 28 bottom, 32 bottom, 33, 34,
38, 42, 43, 47 left, 51, 53, 60, 62, 63 bottom, 64 top, 66,
67 right, 87

SP Creative Design 2, 8, 12, 13 bottom, 16 top, 21 bottom,
25 bottom, 27, 29, 30, 31, 37 top, 39, 40, 46, 48, 49, 50,
54, 55 left, 56, 63 top, 65 bottom, 69 top, 70 top, 72, 73
top, 75, 79 left, 83, 86 top, 85, 88, 89, 91 left, 94, 95

Author's acknowledgements
I would like to extend my sincere thanks to everyone
who worked with me on this book. Most especially, I am
indebted to Trevor Davies for ensuring the project came
my way, my editor Heather Thomas for her help and
encouragement and her colleague Rolando Ugolini for
his superb design skills.

Contents

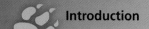

Introduction

Well, what do you want? This is surely a question every cat owner has asked a feline friend at some time or another. So often your cat obviously does want something but his perplexed and exasperated human simply can't work out what that something is!

When all is basically well with his world, the odd misunderstanding between a cat and his owner may be relatively inconsequential. Sadly, however, our cats are increasingly troubled by modern conditions and the environments in which we all find ourselves living. And because we love our cats, enjoy sharing our homes with them, find so much pleasure in caring for them and watching them go about their daily lives, we regard them more than ever as valued household members. It is all too easy therefore to forget that we actually hail from different species and cats have needs and wants of their own, which, unfortunately, do not always coincide with ours. This can cause problems.

Of course, not all unfulfilled feline needs and desires end in problematic behaviours, although a significant number do. We now know that despite the best efforts of their devoted owners, many cats struggle with circumstances over which they have no control and which do not give them what they want and actually need to stay physically fit and healthy and emotionally well balanced. As a responsible owner, don't let your cat become one of them.

The aim of this book is to help you to understand what your cat does and why he does it; in other words, to perceive your cat's world from his perspective, knowing all the factors that have combined to shape his behaviour and all the things he needs from you and his home environment. Once you have mastered the art of feline comprehension, you will be well on the road to being the best as well as the most loving owner your cat could possibly have; and also one who is skilled in giving him every day of his life exactly what he wants.

**SKILL
LEVEL**

Skill 1: Comprehension

Perceiving the world as your cat does is not as easy as it sounds but it is the first essential skill you need to be able to give him everything he wants from life. We share the same basic senses but size differences and anatomical and physiological variations mean that your cat's perception of everyday objects and events is not the same as yours. The way in which he navigates his home and outdoor environments is also different. Understanding the significance of these differences is essential if you are to 'get onto his wavelength' and appreciate his sensitivities – the first steps towards responsible ownership.

Feline senses

To understand your cat's perception of his environment, try getting down onto his level and look at what he sees. Then when you have learnt how his senses differ from yours, you can try to imagine how things actually 'look' from his perspective and you will already be more aware of what he wants.

🔺 **Scent rubbing helps to re-establish bonds between cats and their owners.**

Smell

Your cat lives in a world where scent is supremely important. Cats use it to convey information to each other, label what they lay claim to as theirs and make themselves feel secure. They register environmental changes by minor variations in scent and identify individuals they are familiar with because they smell reassuringly 'normal'. This is why your cat sometimes sniffs you intensely when you return home before rubbing up against your legs to deposit his scents and familiarize your scent profile.

His nose is lined by a membrane that contains odour-detecting cells and is considerably more sensitive and extensive than yours. He also has a vomeronasal organ in the roof of his mouth which enables him to 'taste' some scents, particularly those deposited by other cats. You may sometimes see him intensively sniffing something, then 'grinning' with his mouth open and upper lip pulled back. This Flehmen response is particularly important for entire cats during courtship and mating, but neutered cats also exhibit it. When he goes outside, your cat may sniff the lower branches of nearby shrubs very intently, displaying the typical Flehmen facial expression, then rub up against the object or even spray it with urine. He is registering as much information as possible from the message another cat has left before leaving his own.

Hearing

The size and shape of your cat's ears and their remarkable mobility allows them to move independently of each other in order to collect as much information as possible. The frequency ranges over which cats hear is greater than ours, which allows detection of rodents rustling in the undergrowth and the conversations mice have with each other – something that was essential in the effective hunters from which your cat is descended.

Moving ears

When your cat appears to be resting, you may notice that his ears occasionally twitch or move, perhaps only one of them at a time. He is still aware of what is going on around him in anticipation of something significant happening, such as the approach of a foe or even a potential meal.

○ Something's changed. This cat is learning what.

◁ Cats' mobile ears reflect what's happening around them and how they feel.

9

△ **Constricted pupils indicate arousal and response to bright light.**

△ **Dim lighting, rather than emotion, caused this resting cat's pupils to dilate.**

Sight

Your cat's binocular vision is similar to yours, but the position of his eyes helps to give him a slightly wider field of view. His pupils, despite being less mobile, are more able to dilate and contract in response to light intensity and his emotions. 'Reading' his signals will help you understand how he is feeling in situations where he may not be reacting overtly. This is a very useful skill to cultivate, as it will improve your sensitivity to his mood and help you identify seemingly inconsequential things that he may find stressful.

Detecting colour

Your cat's eyes do contain some of the cells that are needed to detect colour (cones) but as colour has little significance in the feline world it is thought the cat's brain does not process information from them very well. Many of the colourful toys on display in pet stores may well attract you but appeal very little to your cat, especially if they are heavy and slow moving.

This makes sense when we look at the fact that cats are 'programmed' by nature to perceive and respond to fast moving prey more than 2m (6ft) or so away and their sight is adapted for this. Consequently, when something is placed under his nose, your cat will locate it mainly by scent and vibration and what he can hear if it makes a noise. Bear these adaptations in mind when you buy or make toys for him. Although he will have his own favourites, cats generally enjoy toys with prey-like qualities, being small, fast moving, squeaky, shiny and easily manipulated with a paw.

Touch

Like us humans, your cat has receptor cells in his skin to detect touch, pressure, temperature and painful stimuli. Information from them is processed in his brain, which then transmits messages via the nervous system to initiate an appropriate response.

Your cat also has whiskers – specialized hairs around his face and 'wrists', which are vibration and temperature sensitive. The information they convey to his brain about the position of his head and legs relative to air currents and objects in the vicinity helps him to make his way around his environment and perceive the movements of small prey if he is out hunting.

Your cat's whiskers also respond to his emotions, which can help you learn to 'read' his feelings once you get tuned into the way they respond to his various states of arousal or relaxation.

Taste

Your cat has taste buds that allow him to detect bitter, sour and salt flavours, on the tip, back and sides of his tongue. Assessing feline taste preferences is rather difficult, but, as cats are obligate meat eaters, it seems their ability to detect sweetness is limited.

Cats' eyes

A layer of specialized cells, the tapetum lucidum, lines the back of each eye and accounts for cats' eyes typically 'glowing in the dark'. This allows more efficient absorption of light and enables your cat to see better than you in lower level light, but he cannot see in complete darkness.

⬆ This cat's senses are 'on high alert'. Note his dilated pupils and whiskers.

◀ This handsome tabby is calm, confident and relaxed.

Ancestral influences

To understand your cat you must know about the forces that shaped his species' behaviour – these are still strongly felt today. Therefore, as strange as it may seem, looking at his ancestors' world will give you the knowledge you need to be the best owner your cat could have.

🔺 **When in hunting mode, whatever his target, there is no doubting your cat's genetic inheritance.**

Origins still matter

Survival was tough for your cat's forebears. The African wildcat (*Felis silvestris lybica*), from which he descended, inhabits the savannah, where the resources that are essential for life – food and shelter – are hard to come by. Pickings being thin, a solitary lifestyle prevails with individual cats coming together only for mating. The female raises the kittens alone before they disperse at sexual maturity, around six months of age, to make their own way in the world.

Territoriality, an essential trait

The size of each cat's territory plus the number of safe, sheltered areas and rodents, small birds and insects it contains are crucial to survival. Your cat may no longer have to fend for himself but the hunting instinct will remain with him to some degree at least. It reveals itself in play, where the stalking, pouncing and shaking needed to be effective hunters make feline games so entertaining for cat lovers to watch.

Territoriality is revealed in your cat's need to intimately know every inch of his own domain, to patrol it regularly and examine any change that could be significant to him. He may be strongly territorial and need to lay claim to, and defend, a large area or be content to call a relatively small patch his own, but this inherited tendency will be there whatever his background and personality.

Small predators may be prey

Even a predator as successful as this well adapted survivor is still small enough to fall prey to larger, stronger animals and birds. As a result, the wild cat's aim is to move silently

⬢ **Like his wild forebears, a cat feels safest when he is surveying his world from on high.**

Short-term interest

The nature of the wild cat's diet also influences the cat that shares your home. Rodents, the preferred prey of these superbly developed carnivores, provide such a lot of high-energy protein that after the draining burst of activity needed to catch a meal a cat can stop and rest. This activity pattern explains the feline trait of intense interest in something novel and engaging that lasts a very short time – providing new toys that are quickly ignored can be disappointing, but such brief interest does not indicate your effort has been wasted.

around his world attracting as little attention as possible. Plentiful, easily accessible hiding places – lower branches of trees and dark crevices – are therefore essential for survival. Being able to retreat and hide, especially in high and dark areas, such as open cupboards and under your bed, will be just as important to your cat, even if he rarely feels the need to do so.

Night-time activity is safest

The need to remain hidden as much as possible also influenced the animals upon which the original cats depended for food. Rodents are active at night, especially at dawn and dusk, so a nocturnal, crepuscular lifestyle became the norm for their hunters. Your cat may have adapted to sleep through the night or his typical pattern may be problematic, especially when the days are long. If so, make sure he has plenty of exercise and interesting activities. Put out a new plaything, such as a cardboard box, every evening. Never shout or get up to feed him – rewarding his behaviour with food and your attention is likely to intensify it and make a rod for your own back!

⬢ **On a mission, this cat is patrolling his territory.**

SKILL
LEVEL

Where are cats now?

Understanding ancestral influences will set you on the right path, but to really understand your cat you also need to know what impact closer contact with people and other cats has had upon the behaviour of our domestic felines.

Who domesticated whom?

Early farmers realized the value of unobtrusive animals with a talent for catching the rodents that were attracted by their harvest, and cats benefited from readily available food and shelter. This mutual co-operation resulted in shy and solitary felines discovering that developing a relationship with people was advantageous and ended with your cat's satisfying presence in your home.

🔺 **Behavioural adaptability helped cats to gradually become our valued household companions.**

🔻 **These related young cats recognize each other through prior knowledge and scent bonds.**

Cats have adapted but only so far

In contrast to dogs, where for thousands of years humans selected those with behavioural traits they could use to their advantage (hunting, guarding and herding livestock, catching vermin and retrieving game), people have had relatively little influence upon feline reproduction and behaviour – cats have simply gone off and found the mates they wanted. Pedigree cats are the exception, although even some of these felines sometimes sneak out and take care of matters for themselves!

Despite some modifications, your cat's appearance and behaviour do not differ significantly from his original ancestors. However, closer proximity to others, both two- and four-legged, has required some degree of change. Individual characteristics also play a part in forming him. If he has a blue-blooded ancestor, you may identify particular signs, such as a characteristic slender, elegant build and oriental face or the typical Siamese yowl.

Science improves our understanding

Studies involving groups of free-living domestic cats that have congregated around rubbish dumps or live on farms, for example, have shown us that the solitary nature of the original cats has given way to a more sociable lifestyle. However, it is important to understand that these situations in which cats are genuinely part of bonded family groups differ from those where a number of unrelated, un-owned feral cats have been attracted to a reliable source of food. Then the unrelated animals remain self-sufficient individuals, even though access to shelter and food requires them to modify their behaviour or risk confrontation and potentially fatal injury. It is the essentials that are necessary for subsistence that draw them to a particular place where they live together, not a desire for feline company.

Know your cat

Looking at your cat afresh now that you are armed with this information should help you to understand how historical influences have combined with his individual qualities to affect everything he does as well as everything he needs from you, including his home life and his daily routines (see also pages 16–17).

⬆ 'Back off and keep away': this tabby's message could not be more clear.

◀ Each pedigree breed has its own appearance and behavioural characteristics.

Your cat's world view

To understand how your cat sees his world you need to refer to the behaviour of his original ancestor, the modifications made by his species during its association with people, and the more immediate influences that affect his outlook and reactions. Considering how all these factors blend with your individual cat's personality and early experiences will help you as an owner to provide the sort of world that will keep him healthy and contented.

⬆ **The right sort of early experiences are needed for cats to become confident, well balanced family pets.**

Genetic influences

Despite common characteristics that were inherited from their feline forebears, our cats' behaviour is also determined by the sensitivities and behavioural traits of their families, especially their parents. Whereas some feline families are bold or sociable, others are timid. Pedigree cats hail from a smaller, more controlled gene pool than 'ordinary' domestic moggies, so despite personality differences, they tend to display certain characteristic breed behaviours.

Each cat is unique

Ancestral and family inheritance is not the only factor that influences feline character. Every cat, like every owner, is indisputably an individual, and even littermates that look identical may be like 'chalk and cheese'. Your cat's individual traits and sensitivities are also determined by what happened to him in the early weeks of life. Born blind and relatively helpless, to some extent he started

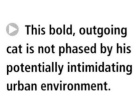

▶ **This bold, outgoing cat is not phased by his potentially intimidating urban environment.**

learning inside his mother's womb. However, nothing at any time throughout his life will compare in significance to the socialization period – between two and seven weeks of age, when exceptionally rapid physical and behavioural development is taking place. Kittens learn about themselves, other individuals – their mother, siblings, people and other pets – and the world around them. If their experiences are limited, they may find later life in a busy domestic environment difficult to cope with.

Anyone raising kittens has a responsibility to ensure that they meet a wide range of kindly people during these early weeks – men, women and children with a variety of ages and appearances. They also need exposure to all the ordinary things that we take for granted, such as noisy household appliances, which may be perceived as scary if they are encountered later in life without the relevant preparatory early experiences.

Later also matters

Sadly, even the best start cannot always prevent things going awry if a cat subsequently has traumatic experiences. Frightening or painful events at the hands of people or other animals can shake his confidence, or frightening incidents involving vehicles, domestic appliances, building work or fireworks may undermine his ability to live happily in a normal bustling home.

Get things right

Hopefully your cat had the right sort of conditions during his early weeks to give him a broad template of people, pets and domestic paraphernalia against which he can compare anything new or potentially frightening. If not, he may find minor changes in your home environment or anyone new intimidating, even when they are kindly disposed towards him. To avoid this happening:

- Take things easy

- Make changes gradually

- Think carefully about how best to handle the situation before deciding to alter anything

- Ask newcomers to 'back off', leaving your cat to adjust and 'make friends' if and when he wants to.

Cat-to-cat relationships

Relationships between cats are often relatively complex. You should never take it for granted that positive associations will endure through thick and thin or that your cat will like another individual just because you want him to do so. Looking at how cats evolved, the circumstances under which they bond to others and the limits of their ability to co-exist contentedly, will help you understand your pet's interactions with other felines, and whether he would be happiest as 'a loner' rather than living in a multi-cat household.

⬭ **Truly bonded cats enjoy each other's company.**

Hostility

Sadly, it is not unusual for cats expected to share a home or live close to each other in the same neighbourhood to become actively hostile, with tension gradually increasing and fights becoming more frequent. They are not being spiteful but simply 'feline' – responding to circumstances beyond their control in the way that nature dictates.

Family matters

Whether a cat views another cat as a friend, foe or competitor will depend upon whether or not they are related, when they first met and how introductions were conducted, past experience and current circumstances. Cats naturally bond with their mother and siblings. They may also bond with others, if they are born into a larger, extended family group or are, for example, introduced to another kitten when very young, whereas sometimes an older, tolerant and accepting adult may form a genuinely positive relationship with a youngster.

Coexistence is not always easy

Many people think that unrelated cats introduced to each other as adults will become firm friends, but it is more likely they will remain individuals living within the same area. In effect, they are two separate feline social groups sharing one space, and within that, even if it is quite small, they will each have their own areas of operation. They will probably shun contact if possible and will find situations where they are forced into close proximity stressful, such as when they need to pass each other in a corridor or hallway, are fed in the same room at the same time or shut up together when their owners are out.

Owner understanding

In any multi-cat home, passive intimidation and active hostility will increase if space is at a premium and there are limited essential facilities, such as high observation

⬆ **Coexistence is not easy: the tension between these two cats is clearly written on their faces.**

⬇ **These cats cope with each other's presence by keeping their distance.**

points, dark and secure hiding places, resting places or latrine sites. Food may be plentiful but, as cats remain solitary hunters even when they live naturally in related groups, eating close to others is a recipe for distress.

Some relatively equable relationships between co-habiting cats, even if they do, to all intents and purposes, pretend that the other one doesn't exist, can deteriorate or aggressively erupt if their owners insist on 'playing happy families'. Expecting cats that share the same home to do everything together just because you love them all is not reasonable. Unless they are truly bonded in their own feline terms, such an arrangement is more likely to cause feline misery than happiness.

Even related siblings may fall out if you do not understand their need to choose when and how much close contact they have. Inadvertently introducing competition into everyday situations, especially at meal times, can be avoided by having different feeding stations in different areas and providing plenty of the essentials of feline life, so each cat always has a choice. Understanding this is key to having a happy feline family.

Cat-to-human relationships

SKILL
LEVEL

By taking into account some fundamental differences between human and feline social needs, and knowing your own cat's background and character, you will be able to understand his reactions, what style of interaction will suit him best and just what he is able to bring to your relationship.

Handling your cat

- Affectionate gestures, such as petting and cuddling, that 'invade his personal space' could stress your cat by making him feel 'trapped'.

- When it comes to getting up close and personal, follow his lead and show your love with appropriate overtures that allow him freedom of choice. Offer a fishing rod or some toys rather than stroking him.

- If he is not a lap cat, respect this and simply enjoy him hanging out with you on his terms.

Cats are not small people

Humans and dogs were drawn to each other because they are both social species that live in groups and find benefits in taking part in cooperative activities. As a result, we all find isolation quite difficult, even threatening, and we tend to enjoy close contact with others. Cats, in general, are different, although some individuals enjoy 'hanging out' with their owners and demanding their attention, especially if they're not getting as much as they would like!

However, when it comes to social interaction the feline norm is the opposite of ours. We go about our normal daily lives and may not see those we care about for hours, days or even longer, intermittently enjoying high-intensity interactions, often at close quarters with affectionate hugs and intimate gestures.

🔺 'Real' lap cats genuinely enjoy the experience of close contact with their owners.

Making predictions

Many experiential and genetic factors influence a cat's personality and behavioural development, so it is impossible without knowing him to predict how an individual cat will view people, particularly whether or not he will enjoy close contact with them.

⬆ **By holding back, this owner is actually encouraging her cat to approach her more closely.**

Cats are very independent, even with those to whom they are well bonded. They like to check in frequently but for short periods, and they may not make any physical contact at all. It is not uncommon for sociable cats to 'say hello' with a quick rub around a person's legs and then retreat to a nearby vantage point, perhaps a sofa or windowsill to observe what everyone is doing. They may remain in the room for a long time or quickly leave, returning later for another 'check in'. Cats, as it were, enjoy high-frequency, low-intensity interactions.

⬇ **Hanging toys have the advantage of giving cats complete control over games.**

Experience and individuality

Temperament also plays an important part in how cats view people and react to them. Some are naturally more sociable than others, but early experiences are crucial. For instance, if kittens are handled only by women during the socialization period, they may find contact with even the most kind, caring men and children unnerving later in life. This is why they need careful handling by as many people as possible, especially during their first few weeks of life. Later good experiences are unlikely to make up for early limited socialization or poor handling, and these cats will probably remain fearful around people and find close contact stressful throughout their adult lives.

Companionship needs

SKILL
LEVEL

Whether or not your cat will enjoy company, human or feline, will depend entirely on him. Foisting the wrong sort of companionship upon cats can be unkind, lead to problematic behaviours and undermine the joy of ownership. If you are concerned that your cat may be lonely, consider his individual history and personal preferences to help you make the right choices for him.

The environment is crucial

When assessing your cat's need for company, be aware of his species' behavioural traits and the importance of the environment in feline terms. Due to their original relatively independent background, cats rely much more on their surroundings to keep them happy than on the presence of other individuals. In this respect, it is often said that 'cats are not small dogs'. Whereas separation and single status can be emotionally uncomfortable for a social species, most cats are happier as 'sole operators', so long as they live in a world that satisfies their general feline needs and particular desires (see pages 48–51).

Where did he come from?

If your cat had a poor start in life he may find anything other than a quiet home life, where he only interacts with

🔻 **Many cats are happiest as 'lone operators'.**

⚠ **An interesting environment is often perceived as being more valuable than the company of other cats.**

a few familiar people and no other cats, too difficult to deal with. In this respect, a chaotic household in which he was neglected and had to fight for regular meals or safe retreats, or suffered constant noise and activity, is just as significant as a kitten having limited socialization and contact with people and other pets.

Remember this when you are deciding with whom your cat should be expected to socialize, especially if you are thinking of getting another cat. Don't expect too much of him nor allow unfamiliar visitors to unwittingly pressurize him by trying to 'make friends'. Enforced attention, particularly from people they don't know well, is often stressful for cats and you should not permit it.

⚠ **Appropriately engaging their 'little grey cells' with prey-like objects keeps cats happily occupied.**

Avoid mistakes

- Working long hours or frequently travelling away from home does not mean your cat will be lonely. Ensure that, from the feline perspective, your home is well appointed and make the most of everyday activities to entertain him.

- You may like being surrounded by cats but that does not mean your cat will enjoy living with the feline companions you choose for him. He may be a lone wolf and find life in a multi-cat home impossible to cope with.

- If he is used to living with one of his family members (his mother or a sibling) or another cat he bonded to when young, think carefully before filling the gap left in your home when you lose his companion. Like you, he will be upset by the loss and changed routines, but don't rush for a replacement. It is his particular companion he will be missing, not just another feline housemate.

SKILL
LEVEL

Skill 2: Communication

Cats communicate with each other – and also with us when we form part of their community – very differently from the way in which we humans do. Failure to recognize this often leads to misunderstandings between cats and people. To really comprehend your cat and what he wants, you need to understand his communication methods. This will help you to make sense of his otherwise rather strange habits and ensure that you provide all he needs in the right places to 'express himself' and keep him happy. Whether he's a chatty cat or moves silently around his world, learning more about how cats communicate will help you see why this makes sense.

Why cats communicate

For a solitary hunter, such as your cat's ancestor, two of the most important aspects of feline life were laying claim to territory and keeping out of trouble. This meant that active defence, with the potential it carried for sustaining a life-threatening injury, was too risky a business to get involved in, unless it was absolutely unavoidable. Therefore, cats developed a range of strategies that were aimed at staking out territorial claims without having any contact with interlopers and potential competitors.

Territorial claims

Ensuring that his domain is firmly labelled as his helps a cat not only to keep other felines at arm's length but also makes him feel more emotionally secure. As a result, cats largely communicate with the aim of increasing the distance between them and are much better at keeping out of each other's way than de-fusing tense situations when confrontation is inevitable.

Familiarity breeds content

Despite the need to maintain a low profile, when cats live in bonded family groups they also need a means of creating a colony or communal identity, recognizing familiar individuals and restoring bonds if tension flares.

◀ Cats only feel able to relax and be emotionally secure when their territory is clearly labelled.

🔵 **Close contact between litter mates facilitates constant scent exchange and communal recognition.**

As scent is their most important sense, unsurprisingly they use odour to identify each other and to maintain their bonds as well as for marking their communal boundaries and important areas within their domain.

If your cat lives with a family member or a bonded feline companion, you may sometimes observe the typical allorubbing familiar cats engage in where they rub up against one another with their heads, flanks and tails, or he may do this to you. Allogrooming is also seen in bonded cats when they relax together or have had a spat – it helps calm them down and restore harmony.

Different messages

As he goes about his daily life you will observe your cat constantly using scent to send out messages and receive information from other cats and objects, such as your clothes when you return home. He will also use visual signals, body language and the rough marks he makes on trees, scratching posts, or, if you are unlucky and unaware of how to cater for his natural behaviour, your furniture!

🔻 **Owners generally find facial and flank rubbing the most acceptable feline marking behaviour'.**

Scent signals

SKILL
LEVEL

It is difficult for us, with our relatively poor sense of smell, to understand just how significant odour is in the feline world. It is the most important aspect of communication for your cat, and he will be acutely aware of even the most minor changes in the scent profile of the objects and individuals around him. Once you understand not only the importance of scent as a communication tool but also how your cat uses it, watching him go about his everyday business will be ever more fascinating from your perspective. It will also be extremely informative when you are trying to work out how he feels about his environment, things, people and other cats and pets.

🔺 Sending and receiving scent messages is crucially important to cats, especially when the local feline population is high.

Scent as a form of communication

The value of odour molecules as an effective signal is that once they have been deposited somewhere although the signaller can leave the scene, his scent signal will still continue to work on his behalf for a while at least.

Pheromones are naturally occurring scents that animals use to send and receive vital information. For example, they can announce an individual's presence, his sex, age, health and emotional state and whether or not he is a friend or a foe. Another important function they have in the natural world is communicating information with regard to breeding status.

Pet cats are generally neutered, so this is not normally an issue. Nevertheless, your cat will constantly employ scent both to navigate his way around his environment and to maintain himself on an even keel emotionally.

Rubbing and bunting

Special skin glands in specific areas of your cat's body – they are located around his mouth, face, between his paw pads, flanks and tail – produce oily secretions containing pheromones (see above). This is why you will sometimes see telltale greasy marks at cat height where he has been rubbing his face and flanks around the furniture or pristine white paintwork. Outside you may see him doing a similar thing with garden furniture, fencing, plant pots, and shrubs.

Scratching and 'stropping'

Your cat will also send out an auditory signal and leave visual territorial marks in the form of a roughened surface when he deposits the scents from between his foot pads while 'stropping'. You may see him stretching up as high as he can reach against vertical surfaces, such as a tree or fence post outside, his scratching post at home or, if you are unlucky, your furniture, door frame or wallpaper. The majority of cats favour stropping in this way but he may be one of those that prefers horizontal surfaces, at least in some places – your carpets may bear the telltale scars, if so.

⬤ **Scratching is natural, so ensure that your cat has suitable facilities to avoid indoor destruction.**

Your cat is also likely to bunt, as this facial rubbing is called, and rub with his flanks cats and other pets with whom he is well bonded as well as your legs when he has not seen you for a while. The intensity of this behaviour will increase if you have been especially vigorous with your household cleaning and obliterated his reassuring environmental marks or you return home smelling a bit strange from his point of view.

▶ Indoor-only cats can still be affected by other cats in their vicinity, even outside.

🔻 Cats often respond to territorial incursions with intense urinary marking.

Urine marking

People often assume that urine marking is only seen with mating behaviour in unneutered cats, but males and females, and even neutered cats, label their territory with urine. Your cat may adopt the classic stance associated with urinary spraying, backing up against a vertical surface and raising his quivering tail while paddling with his back feet before squirting a small amount of urine. If he sniffed intently first, it is likely that another cat has been there before him and the volume of urine applied may be increased. Some cats do not always adopt this pose, and a pool of urine can be used as a signal rather then simply indicating a cat's need to relieve himself.

'Middening'

Using urine as a marker is not the only way in which some cats employ excreta to indicate their presence. You may find faeces in exposed locations, such as the middle of a lawn, a garage roof or on top of a wall, rather than being discretely buried in private areas, as is the norm. This type of marking is less common than other methods but it is especially prevalent where there is a dense feline population and territorial tensions are running high.

Nothing lasts forever

Despite your cat's best efforts, his odorous messages will decay with time and temperature, although, initially, heating the scent secretions may help to broadcast the information they contain over a wider area. This explains why he repeats the exercise, often quite frequently. The intensity of his marking activities will also increase if he is troubled or facing environmental challenges.

Territorial divisions matter

How and why cats communicate is not the only issue when it comes to giving your cat what he wants. What happens where also matters and may affect your home if you don't get things right when providing facilities. He will divide his world into a core area, home range and hunting range, probably sharing the latter with other cats, especially in areas with a high feline population. Unless you have a well-bonded multi-cat group, his core, where he eats, sleeps, plays and rests, will be entirely his own. He will bunt and rub against things and individuals and leave more intense signals by stropping in significant places – entrances and exits or doorways and 'pathways'. This behaviour may increase in frequency and intensity if he feels stressed. Characteristically, urine is used to mark the periphery of his home range, which acts as a buffer between him and other cats. Spraying also 'labels' trails and significant objects within his territory.

Territorial issues

Naturally, males roam further afield and they have larger territories than females. Neutered cats tend to have smaller territories than their entire counterparts, and what space any individual can claim will be determined by sex, personality, how many other cats live nearby and how territorial they are.

◀ **Cats adopt a squatting posture when urinating, unlike their characteristic spraying stance.**

Vocal signals and cat calls

SKILL LEVEL

Historically, cats made little noise because, except when searching for mates, they needed to keep a low profile and draw as little attention to themselves as possible. You may find that your cat is chattier than his forebears or he may retain a fairly low-key vocal repertoire – a lot will depend upon his personality, early experiences and how much you talk to him. Either way, it will help you to understand him if you know when and how cats use vocalization.

🔺 Many cats live relatively low-key silent lives.

Kittenhood

Purring often has a calming effect upon people and we generally associate it with quietly contented cats. In a natural situation it functions with a subdued chirrup as a means by which a female communicates with her offspring when she returns to the nest. Kittens utilize purring to indicate their distress or attract their mother's attention, without making undue noise that incites trouble, if they become separated from her and their littermates. Both mother and offspring also purr when she is suckling them. Otherwise, apart from mating calls and face-to-face meetings that end in the growling, hissing and yowling of aggression, vocalization generally plays little part in feline communication.

▶ Even unseen, some oriental cats can still be identified by their voices.

Chatting with your cat

Over time, closer association with humans has meant that many cats have become more vocal, and some have even developed a range of vocal skills for successfully manipulating their owners! Few cat lovers can resist the charm of their cat trotting to greet them, tail up and chirruping away. Rewarded with human attention, this vocal habit quickly becomes established and mutually satisfying for both the owner and the cat.

Chat – don't treat

Delightful as chats with our cats can be, if yours is a vocal cat, do not imagine that whenever he comes to greet you or find out what you're doing, he must be hungry. Many veterinarians are now concerned that these misunderstandings contribute significantly to the 'epidemic' of feline obesity. Cats are often able to predict the actions of owners to whom they are well bonded and who control their food supplies. They realize that people often go to the kitchen first when they arrive home to make a drink or cook a meal, and they turn up on cue. Their owners conclude they are hungry when they may just want to touch base after a period of separation. Food that is offered will be obligingly eaten, and before long a sociable, chatty cat will be consuming many more calories per day then he actually needs. If your cat falls into this category, try substituting a game, a short petting session or a comforting 'conversation' instead.

Vocal breeds

Almost any cat with the right temperament and background may turn into a feline chatterbox, especially if his owner is similarly vocal. However, certain pedigrees, such as the sociable Siamese and Burmese, are renowned for their 'voices'. Their rather raucous cries also tend to be passed down to mixed blood offspring. If you inadvertently encourage too much of this vocalization, things can get out of hand and you may long for some peace and quiet.

Even really young kittens are quite vocal but as they mature and grow into adults they use their voices less.

Body language

SKILL
LEVEL

When scent and other feline signals fail to keep cats at a distance or they are misunderstood by people, cats have little choice but to resort to more overt gestures to make their feelings known. Once you have learnt what your cat is trying to 'say' with his visual signals, you will be better able to 'read' him and his emotions. Study his body language to discover how he feels.

Body posture and tail position

You can tell a lot by your cat's muscle tone (how relaxed or tense he looks), his silhouette and how he holds and moves his tail. Never try to analyze one individual aspect of his visual repertoire – this can be misleading. It is very important to take everything you see into consideration when you are working out exactly what he is trying to convey to you or other pets or people.

Facial expression

Some feline facial signs are quite subtle whereas others are obvious. There is little doubt that a snarling, growling or hissing cat is highly aroused and may attack if approached. However, observing his ears and eyes and how his mobile whiskers move in response to his changing emotional state will also tell you a lot.

△ Although he is fairly relaxed, something has caught this cat's attention.

▷ Stretching his whole body in readiness for action, this cat is on the move.

This 'tiny tiger' is becoming defensive.

What your cat's eyes tell you

Bright light will cause your cat's pupils to constrict and his eyelids to narrow as will the heightened emotional state that makes him assertive or aggressive in defence of something he values. Low-intensity light will make his eyes look enormous due to his widely opened eyelids and dilated pupils. Narrower lids with black, dilated pupils are a sign of apprehension or fear, while droopy eyelids signify relaxation and drowsiness.

Ear position

Look out for the following:

- Alert but unruffled cat – forward-facing pricked up ears
- Anxious or fearful cat – stiff, flattened and lowered ears
- A threatened cat trying to protect his ears will flatten them against his head
- Aggression is indicated by slightly flattened, back-turned ears and an assertive body posture.

Whiskers

A cat's whiskers will:

- Droop slightly when he relaxes
- Extend sideways and downwards when he is interested and confident
- Be pulled backwards to protect them if he feels anxious or threatened
- Stiffen and point forwards when he is hunting, playing or trying to intimidate.

SKILL
LEVEL

Interpreting your cat's signals

Once you know how your cat views his world, communicates with others and makes himself feel safe you can identify the purpose behind many of his daily rituals. Getting used to his normal reactions and territorial marking behaviours will help you to understand his signals and how he feels about his world as well as interpreting the meaning behind changes in his behavioural patterns. This is an extremely important step towards being a sensitive owner.

Changed patterns

When cats are feeling stressed, either by life in general or a specific environmental or social challenge, they may exhibit frenzied sniffing and marking. Their dedicated and intense examination indicates they are trying to gain every possible scrap of olfactory information about something or someone before marking them well, in an attempt to make themselves feel more in control. Sadly, some cats find life so difficult, stressful or frustrating that they even bring urine marking and middening indoors.

Get to know normal behaviour

When cats are happy and relaxed they greet familiar people and pets they regard as part of their social group with bunting and rubbing. Cats that are bonded enjoy nose-to-nose greetings before going round behind each other and sniffing under their tails. They also mark a range of environmental items when they are out and

◉ Properly meeting your cat's natural needs can help to avoid destruction.

🔺 **This cat has really relaxed because his home feels comfortable, predictable and secure.**

Scratching posts

Many scratching posts are too short and wobbly. Cats prefer natural fibre coverings with a vertical thread down which they can rake their claws. This keeps their nails in trim and leaves good visual and scent signals while producing a clear auditory warning of their presence. Provide enough tall posts or horizontal scratching pads, which must be stable with a suitable covering. Experiment with location, but these essential facilities need to be in places your cat regards as important or where he feels vulnerable: by outside doors, his cat flap, entrances to rooms, on landings and at the bottom of your staircase.

about, first encounter them, have just woken, or the objects are new. This behaviour can often be relatively intense and possibly prolonged, but the overall picture the cat paints is one of contented interest.

Read the emotional signals

Get used to reading your cat's emotional signals – and acting appropriately. If he is disturbed, try to make things less challenging from his perspective. His world can be a predictable and comfortable place if you, as an owner, provide plenty of dedicated facilities for him to mark his territory. Position the things he requires where he needs them, not where they are convenient for you.

If you do not have enough scratching posts, or you position them where they have no meaning in feline society, they will be of no possible value to your cat. He may even start to damage your furniture instead, causing increased stress all round. Understanding what he needs and getting his world right will help him cope with the everyday challenges that life throws at him.

Playing your part well

Visitors who don't like cats may often appear to be feline magnets with cats flocking to their sides. This seems less perplexing when we examine the way in which cats interact with others and the style of interaction they prefer people to adopt. Learn to 'read' your cat well and you will become adept at eliminating inadvertently stressful interactions that well-meaning cat-loving people often unwittingly impose on the cats they love.

⬭ **Intense sniffing often demonstrates that a cat is disturbed by introducing new 'foreign' odours.**

Quiet arousal or acquiescence?

Many sociable, well-socialized cats enjoy being around people – their owners in particular – but this does not necessarily mean they are 'lap cats' or wish to snuggle up beside them. They may simply want to hang out in the same space, part of the group but without having any close contact. This can be disappointing but cats have a fundamental need to control interactions. So 'trapping a cat into a cuddle' can undermine an otherwise fulfilling relationship and even lead to aggression and injury.

A major mistake people make is to think that because a cat does not struggle and lash out he is enjoying their affectionate hands-on petting. They often overlook his flicking tail tip, dilating pupils and tense muscles. This increasing arousal may boil over but many cats are not temperamentally predisposed to be overtly aggressive. Therefore, instead of making their displeasure obvious, they simply internalize this stress or avoid future contact.

Cats and people are different

Cats and people have different social styles. We are truly social and enjoy company. We put a lot of effort into breaking down social barriers with communication systems that reflect this. When we see people we care about, or want to get to know, our facial expressions as well as our body language naturally draw us closer together with hands-on gestures invading the other party's personal space to express our affection.

Unfortunately, our cats aren't like us. When they have a relationship with people, they like to check in quite often, 'say a quick hello' by chatting, bunting and rubbing, and maybe join in with activities – have a game with a computer mouse or pen, or sit around for a while – before wandering off. Even the friendliest cat has a different interactional pattern from us.

Respect your cat's individuality

By hanging back and quietly observing your cat's natural 'style' you can learn to read him well. Always take your lead from him – stop petting him the minute he tenses, his tail starts flicking or his pupils dilate. Better still, get so good at interpreting his emotional state that you stop your affectionate interaction before he becomes this aroused. High arousal increases the risk of an aggressive outburst in an otherwise sociable and people-friendly cat.

Also, be sure to give him the option as to whether to come close or not by offering a fishing rod toy or some feathers on a stick. Keep several of these toys in different locations around the house, so that something is always available and to hand. Don't be disappointed if your cat does not want to interact; many owners find that their relationship with their cat improves when they adopt this approach, and their pet chooses to spend more time in closer contact with them once they learn to back off.

⬆ **Early gentle handling is an essential element in socializing kittens.**

Visitors

Ask visitors to adopt a more laid-back, non-interactive approach. And no matter how much you love your cat and proud you are to have such a wonderful pet, never pick him up and thrust him towards a guest. Such eye-to-eye confrontation and lack of choice is unkind and a recipe for feline distress. Instead, allow him to come forward to engage with a toy, only if and when he chooses to do so.

◀ **This cuddle is well intentioned but obviously stressful for the cat.**

SKILL
LEVEL

Skill 3: Understanding

Whatever your cat's age, being familiar with his general condition and usual patterns of behaviour will help you to identify potentially significant changes and quickly react to signs of illness or injury. Never delay in consulting your vet, even just for advice, if you spot something odd or your cat seems under the weather. There's more to good health, however, than physical fitness, and understanding the influence of behavioural issues upon your cat will also greatly assist you in safeguarding his emotional and mental wellbeing.

Keep your cat healthy and happy

△ **Bright eyed and bushy tailed, this youthful cat is in peak condition.**

Good and effective health care for your cat requires a co-operative effort between you and your veterinary team. Make a point of regularly checking him over each day, running your hands through his fur and over his legs and tail, looking for anything that is amiss as well as monitoring all the other signs that are associated with good health.

Behavioural signs of good health
- Bright and active (this will vary with age), not lethargic and dull, confused or disoriented.
- Interested in what's happening in the home, not withdrawn and listless.
- Normal appetite, not anorexic (off food) or eating voraciously.
- No evidence of coughing, sneezing, vomiting, diarrhoea or constipation.
- Drinking normally with no unaccountable thirst increase (for example, due to hot weather or the heating being turned on in winter).
- Breathing normal, not laboured.
- Normal pattern of urination – no increased frequency, signs of distress when urinating or haematuria (bloody urine).
- No stiffness, reluctance to jump up, lameness or signs of pain when you pet your cat or he tries to eat.

Signs of good health

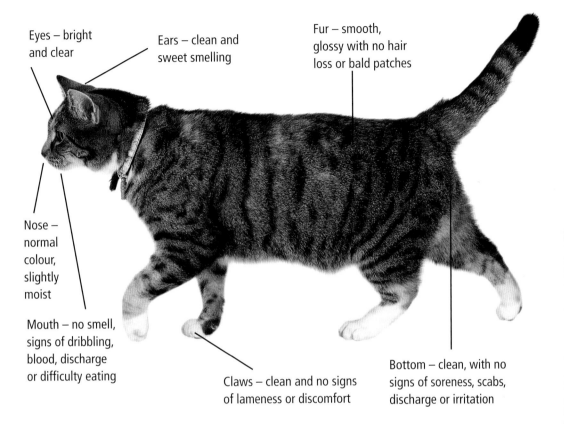

Eyes – bright and clear

Ears – clean and sweet smelling

Fur – smooth, glossy with no hair loss or bald patches

Nose – normal colour, slightly moist

Mouth – no smell, signs of dribbling, blood, discharge or difficulty eating

Claws – clean and no signs of lameness or discomfort

Bottom – clean, with no signs of soreness, scabs, discharge or irritation

Danger signals

Watch out for the following telltale signs of poor health:

- Eyes – dull or signs of redness, swelling, discharges or soreness associated with the surrounding skin
- Nose – dry, crusty, running or any sign of nosebleeds
- Mouth – dribbling, blood, smelly discharge or difficulty eating
- Ears – dirty or smelly discharge, swelling or wounds
- Fur and skin – dull and clumping, patches of hair loss, signs of irritation, scratching
- Bottom – sore, dirty, with signs of discomfort or irritation
- Claws – too long, smelly or bleeding, associated with swollen digits or paw and signs of lameness.

Wounds

Wounds of any sort require professional advice and may need immediate veterinary attention unless you are sure that they are superficial and your cat is otherwise unharmed – you must never ignore any sign that all is not well with your cat.

39

Preventive health care

Whatever your cat's age, you must attend to some routine medical issues. Consult your vet to put the best possible health care regime in place, and check whether and when you should change this as your cat grows older. It is always advisable to consult the professionals for this sort of advice and never to take any chances with your pet's health and physical welfare.

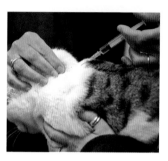

🔺 **Always select a feline-friendly clinic for your cat.**

Vaccination

Your veterinarian will be able to advise you about the recommended appropriate vaccination regime, bearing in mind your cat's lifestyle and medical history. Routine vaccination is recommended against the following:
- Feline parvovirus (panleukopenia or feline infectious enteritis)
- Feline herpesvirus (FHV-1) and feline calicivirus (FCV), the cause of many cases of cat flu
- Feline leukaemia virus (FeLV).

In certain circumstances, your vet may also recommend vaccinating against the following: feline chlamydophilosis, feline bordetellosis and rabies.

Note: Vaccines are available in some countries against feline immunodeficiency virus (FIV) and feline infectious peritonitis virus (FIP).

🔺 **Your vet will show you how to administer routine care, such as flea products.**

Parasite control

Regular preventive treatment against the range of internal and external parasites that can affect cats is an important aspect of keeping your cat healthy, especially if he has access to all areas of your home. Although many over-the-counter products are available, your vet will be best placed to ensure that you use safe and really effective ones as well as advising you about environmental measures to keep your pet and home free from fleas. The parasites you need to consider will depend on where you live – some countries are free from problems that can be life threatening in other parts of the world. Check this out with your vet.

External parasites

Fleas – these tiny, dark brown insects can be observed scurrying through a cat's fur, but are often identified by their droppings, which resemble black sooty grit.

Ticks – vary in colour from slate grey to brown and tend to be found in rough grass. Once attached, they swell up before dropping off when full of blood. They can cause irritation where they attach; always seek professional advice rather than trying to remove them yourself.

Mites – cats suffering from ear mites generally show signs of irritation and their ears produce dark, crusty wax.

Cheyletiella – cats affected by these less common mites generally produce significant quantities of dandruff.

Harvest mites – a seasonal problem in certain regions; can cause irritation in the ear flaps and between the toes.

Internal parasites

Roundworms, tapeworms, lungworms, whipworms, hookworms, heartworms, Giardia and Toxoplasma. Your vet will advise on how to treat and prevent these.

Monitor your cat's food intake carefully to check that his appetite is healthy, not excessive or reduced.

Nutrition

Your cat's physical and mental health depends on a well-balanced diet that supplies all the nutrients he needs. Species, age and weight will influence his nutritional requirements, which may change if he is affected by a specific medical condition. Cats are obligate carnivores. They must eat meat because they cannot get all the essential nutrients it contains, such as the amino acid taurine, from any other source. It is difficult to ensure that home-produced diets contain everything they need to grow when young and stay healthy when mature. Never give your cat dog food – they have different nutritional requirements.

Neutering

SKILL
LEVEL

Neutering, or de-sexing as it is often called, is commonly recommended as the most reliable means of feline population control. The operation must always be carried out by veterinary professionals. In males, it generally involves castration – the removal of the testes – whereas female cats are routinely spayed by ovario-hysterectomy, a somewhat more invasive abdominal operation whereby both ovaries and the uterus are surgically removed.

Timing is important

The best time to neuter cats is widely debated but ensuring that it is performed before the onset of sexual maturity helps to prevent unwanted early pregnancies and the associated risks of sexually transmitted diseases. Unfortunately, although puberty tends to occur at around six months, this is variable with some females and pedigree cats coming into season (or heat) before then. Your vet is the best person to advise as to when the operation should be ideally performed.

Physical effects

The effects of early neutering are not so noticeable in the female, but entire tom cats generally become larger and stockier compared with males that were neutered when they were younger. Their large, broad cheeks tend to be

⚠ **Battle-scarred entire tom cats quickly lose their handsome good looks.**

▶ **Neutering has other benefits in addition to population control.**

◀ Kittens are great fun but they can quickly wear their poor mothers out.

their most noticeable secondary sexual characteristic, although their active hormones result in pungent smelling urine, which many people find offensive. The reduction of these features is another good reason for neutering male cats relatively early.

Behavioural consequences

Entire cats in search of partners advertise their presence and willingness to mate with urinary spraying, while males roam far and wide in their quest for receptive females. This territorial expansion often brings them into conflict with cats whose territories they invade and other entire males they regard as rivals. Therefore, in addition to the characteristic caterwauling by which they announce their availability, frequent fights are also associated with the mating season. Sadly, this behavioural pattern, often involving unwary wandering over significant distances, does put un-neutered tom cats at risk of being involved in road traffic accidents as well as increasing their chances of developing post-fight infections – sexually active females naturally defend their nest sites and resources. One of the benefits of early neutering in both sexes tends to be that our cats remain more youthful in behavioural terms, which many people believe makes them more satisfactory, docile companions.

Monitoring weight

Both sexes, when neutered, have a tendency to convert food to fat more effectively now that they are no longer able to reproduce. Therefore it is important to monitor your cat's weight carefully and also to moderate his calorie intake if necessary. A number of specially formulated diets are now widely available, catering for the nutritional requirements of neutered cats.

Mental and behavioural health

SKILL LEVEL

The link between physical, mental and emotional health is well established and vets are now identifying an increasing number of feline disease conditions, such as feline idiopathic cystitis (FIC) and even obesity, where psychological and emotional influences play a significant role. Consequently, being sensitive and understanding are as important in keeping your cat healthy as is taking care of his physical needs. Once you are aware of what this means in feline terms you will be better able to provide a home environment and management regimes that cater for his emotional equilibrium and physical fitness.

⚠ **Never trap your cat in a situation that stresses him. If he does become stressed, handle him with care to help him quickly recover.**

Protective responses

No one likes to feel that their world is spinning out of their control and cats are no exception. Their forebears survived by knowing every inch of the territory they inhabited and monitoring even small variations within it. They also benefitted from the fear response without which none of us would last very long. It is the strong emotional response that helps us all avoid falling foul of harmful situations or activities. At such times of potential threat our heightened physiological arousal, or stress response, governed by the sympathetic nervous system, readies all the body's systems for flight or fight. In evolutionary terms, this is an adaptive mechanism without which our cats would have died out long ago.

Equilibrium

Natural caution and a stress response that readies us for appropriate action have kept people and cats safe and allowed both species to thrive. But this is only possible when in between 'action stations' alerts, we can relax and regain a state of equilibrium or homeostasis. Controlled by the parasympathetic nervous system, this allows a period of calm whereby a cat or person can regain their confidence and reassess their situation.

Anxiety

Inextricably linked with fear is anxiety, another emotional state that quickly attaches, often without anyone being aware of it, to signs and actions that something potentially unpleasant is going to happen. This is why so many cats panic when their owner approaches the cupboard where their carrier is stowed. They become anxious because on previous occasions these actions have spelled out bad news in their terms.

Chronic stress is harmful

Unfortunately, when conditions are such that our cats cannot regain their equilibrium in between periods of high arousal, stress becomes on-going, compromising the immune system, memory and learning, and adversely affecting their physical and emotional wellbeing. This is why it is so important to understand your cat and to make sure that your home provides the necessary conditions for him to adopt the natural coping strategies with which nature equipped him.

▶ **Being higher feels safer, so be adaptable where your cat is concerned and let him eat where he feels safe.**

Your cat's needs

Sense of security – this will come from having a predictable world. Take care if you are making changes to the layout of your home – don't do anything radical if your cat is already upset by other issues.

Sense of control – having a choice of calming facilities will help to give your cat the perception that he is in charge of events – not the other way around.

Complexity – plenty of interesting things to do will keep your cat occupied and will give him the chance to use his 'little grey cells'.

Novelty – boredom does cats no good at all. Find interesting new activities and toys to keep him busy and occupied.

Social contact – under your cat's control and available when he wants it rather than imposed upon him by you when he doesn't.

◀ **Cats and dogs can sometimes become firm friends, but their owners must be realistic.**

Lifestyle matters

SKILL
LEVEL

Your cat's lifestyle may well be determined as much by your home situation and where you live as anything else. Many owners are now driven by fears for the physical safety of their pets to keep them indoors, or they live in circumstances, such as apartment blocks, where it is not feasible to provide outdoor access. However, many others, with more space in less hazardous environments, allow their cats a more natural, free-roaming lifestyle.

⬆ **Good equipment appropriately placed is essential, whatever your cat's age and lifestyle.**

What's right for your cat?

Carefully weighing up all the relevant issues before deciding what is in the best interests of your cat will ensure that whatever management regime you decide upon will be fair and appropriate, given his age, health, personality and life experience.

Indoor-only versus outdoor-access

Undoubtedly the big wide world can be a dangerous place. Cats can sadly come to grief in a variety of situations: road accidents, drowning, poisoning from plants or hazardous substances and serious injuries from falls are all potential risks when they go out and about. And the higher the local feline population, the greater the risk of cats being stressed by more assertive individuals or becoming involved in fights when close proximity undermines their ability to avoid each other. Potential exposure to infectious disease is also a problem where cats inhabit the same area or meet face to face.

For some physically frail, timid or temperamentally unadventurous cats, an indoor-only existence may actually be preferable, because of the greater stress they experience when facing the more unpredictable outside environment. Indeed, some very easily intimidated felines choose to remain indoors even when they have access to a garden and could explore outside.

Life indoors is not all rosy

Unfortunately, whatever a cat's lifestyle there is no such thing as a risk-free existence. Accidents still happen at home and, unless owners are very clued up, cats can fall foul of poisonous plants and toxic everyday household substances. Physical risk must always be weighed against the advantages that greater freedom of movement provides and the benefits of exploration and exercise that accompany trips outdoors. The opportunities to escape outside from household pressures, produce stress-relieving endorphins by indulging in an intense physical activity, and have somewhere to 'chill out' are especially important when space is limited and/or cats have to share their indoor environments with people and pets they find it difficult to tolerate for whatever reason.

Territorial invasion

Don't assume that because cats don't venture outside they are not affected by other felines. Every time you open a window or exit door, the chances are that the scent of other cats floats into your pet's core area. Such territorial invasion can sometimes be highly stressful to a sensitive or already aroused feline. The risk of overt territorial invasion is, of course, greater when a resident cat has a regular cat flap rather than one operated in response to his identifying microchip or controlled by a magnet he wears on his collar.

⬆ **Outdoor adventures can be risky but many cats thrive on them and enjoy the freedom to roam outside.**

⬆ **Keeping indoor cats healthy and happy requires considerable effort on the part of their owners.**

Making your home 'cat friendly'

SKILL
LEVEL

Ultimately, whatever lifestyle you decide is best for your cat, you must provide everything he needs to keep him physically fit, mentally active and emotionally comfortable. However, this is more of a challenge than it might appear now that we favour minimally furnished, clutter-free homes and gardens.

Choosing dedicated facilities

Combine your knowledge of normal feline behaviour with what you know about your cat's personality to choose the dedicated facilities he needs and where to place them, considering his need to retreat and readjust in a suitable sanctuary and put distance between himself and anyone or anything scary. Considering how his ancestral inheritance, familial traits, individuality and experience influence his perceptions and requirements is extremely important in furnishing his ideal home.

🔺 **Sensitivity is essential where your cat's latrine is concerned – some cats prefer covered trays.**

Creating a cat-friendly home

View your home – your cat's world – through his eyes. Get down to his height and try to imagine just how daunting certain areas are, or comforting it is to hide under a bed or inside a wardrobe. Use your knowledge

🔺 **Growing cat-friendly plants can enrich the life of your indoor cat. Check them out at your local pet shop.**

🔵 **Sharing yours may be fun but your cat needs his own beds, located in safe areas around your home.**

and examine your environment in this way to ensure that your home really suits him and that it provides everything he needs to keep him happy and well balanced.

Your home matters

If your cat has access to outdoors don't think that what you do inside the house doesn't really matter. It does, even if he only graces you with his presence for relatively short periods. If he has an indoor-only lifestyle, inside the house is his whole world and what could be more important than that? And even if you only keep him in overnight or when you go away, you still have to treat him as an indoor-only cat in terms of the feline-friendly environment you provide.

Make adequate provision

All your cat's activities must be catered for (see panel right). Make sure you know what they are and what you need to give him. Don't forget the feline need to be high up. Effectively using 3D space is a great way of avoiding having to sacrifice precious floor space.

Cat-friendly home

Hiding – lots of boxes, bags, tunnels, cupboards, spaces under beds and on top of wardrobes and shelves.

Running, jumping, climbing, exploring, observing – encourage games down hallways and up stairs; provide climbing frames and activity centres (commercial or home-made).

Resting, sleeping – several beds and comfortable areas. If your cat doesn't like what you provide, you can experiment with design and location – don't give up.

Marking – his choice of scratching posts/pads where they make sense to him!

Hunting and playing – lots of interesting toys that he really likes to play with.

Eating and drinking – learn what to do in Skill 5.

Eliminating – adequate toilet facilities indoors (litter tray) or outdoor latrines (see pages 50–51).

Socializing – what suits your cat and when it suits him rather than you!

Outdoors – not 'out of mind'

Many owners mistakenly believe that if their cats go outside they will find plenty of shelter, interesting things to do and accessible suitable latrine areas, but this is often not the case. In towns and many rural housing developments, homes are packed closely together with few mature tees to provide climbing opportunities, observation points and shelter. Patios, decking and open lawns are intimidating for cats, making it more difficult to find suitable toileting areas. Considering what your cat has to face when he goes outdoors is the first step towards making this outside space a mutually satisfying place to be.

Nature knows best

An outdoor 'cat heaven' is any area where a cat can move about without feeling exposed to the view of anyone. Being able to make his way from the shelter of one bush to another, climb a tree for a good view of what's going on, sit up there and doze while feeling safe, or slumber in a sunny patch of warm, bare earth sheltered from the wind, is ideal. And when the weather is bad, or he's not feeling brave, being able to quickly find a suitable area of dry, light soil that is screened by natural vegetation to use as a latrine site is just what he needs.

If your garden does not naturally meet these needs, you don't have to spend a lot of money to improve it. Simply moving large plant pots or garden furniture to strategic locations can create refuges and hiding places,

🔻 **Bold cats enjoy the outdoor life, but you must ensure that your garden is free from hazards.**

Indoor-only cats

If your cat is an indoor-only chap don't think outdoors doesn't matter. Is it possible to install an outdoor run he can access when he chooses via a window or cat flap? If so, don't forget to make this a great place to be with everything he needs to keep him happy. Even if you can't give him outdoor access, feeding the birds, hanging mobiles or glittering garden ornaments outside could provide him with hours of fascinating entertainment.

◯ Be sure to supervise all outdoor excursions until you are confident that your kitten can cope alone.

while old shelf units and garden tables can become elevated vantage points to help him check up on what's going on in his territory, and feel safe while he's doing it.

The right toilet facilities are essential

When it comes to eliminating, cats are clean and private. We have a responsibility to ensure they have suitable conditions, indoors or outside. They like:

- Material that is light and easy to dig a hole in and to cover their excreta
- Secluded areas, near the edge of their territory, where they are not on view. Indoor latrines should not be near their food or in busy rooms, where it is difficult or frightening to access them. Outside they need to be screened from view by bushes, fences or garden walls
- Granular, clumping substrates, such as sand, loamy soil and cat litter – cats' preferences are formed early in life, so it helps to know what a kitten's mother taught him was appropriate toileting behaviour
- Clean latrines – make sure you regularly attend to your cat's litter tray or outdoor area, especially if you have a small garden and it is often used.

◯ Getting 'back to nature' certainly suits this confident cat as he explores and surveys his surroundings.

Skill 4: Motivation

SKILL
LEVEL

You will quickly learn to look behind anything your cat does that seems unusual, intriguing or frankly annoying for the motivation that is driving his behaviour. There is always a reason for feline actions, and this will generally become clear when you interpret them in the light of his species' requirements, genetic inheritance and the early experiences that shaped him. You must also understand the influence that evolutionary survival strategies have on his everyday life. By becoming a more perceptive owner, you can pick up the signs and know how to help him in a positive way if he is struggling to cope.

Focusing on emotional reactions

Your cat will be more at ease with life if he is bold and confident and had a sufficiently wide range of experiences as a kitten to regard his environment with equanimity. Even so, he is likely to 'spook' from time to time, such as when someone drops a cup on the floor or he hears a dog bark. His fearful reaction will activate the stress response that readies him for action if something truly threatening is taking place (see pages 44–45).

⬢ This tabby's unblinking stare is unsettling his companion. You can learn a lot about your cat's emotional state by observing him closely and how he reacts.

 Early life and individuality will determine what a cat finds stressful or frightening.

Survival strategies and learning

What your cat perceives as threatening may actually be something ordinary that you take for granted. Don't underestimate his heightened emotional state; a lot will depend upon his parents' temperaments, his personal/individual sensitivities and early experiences. Be especially sensitive if he is timid or has come from a background where he wasn't able to learn in a non-scary way about the normal activities of home life. The pinging of the microwave, a tumble dryer on a spin cycle or noisy family interactions may all be normal for you, but such everyday events may seem unpredictable and frightening to some cats. Knowing this will enable you to cater for your cat's sensitivities and take the appropriate action.

Feline responses

When your cat meets something new, puzzling or potentially frightening he can respond in one of four ways (the 4F's). A threat in feline terms is often perceived rather than real, making us insensitive if we don't view things from a cat's perspective.

Flight – most cats will retreat if they can; a short distance to reassess or a high-speed flight depends on how they perceive a situation. Never stop a fleeing cat: you may get hurt and he will be more stressed.

Fight – when the challenge feels overwhelming or flight is impossible, a cat may feel this is his only choice. Back off at once.

Faff/fiddle about – displacement activity is a sign of emotional conflict. Watch for your cat grooming in odd situations to calm himself down. React quickly to resolve problems; don't add pressure by petting him or making him 'face his fear'.

Freeze – this indicates real distress. Provide a calm, dark environment and radically review the situation.

SKILL
LEVEL

Early influences

The more you know about the conditions in which your cat was raised, what happened to him and who he met during his first few weeks of life, the easier it will be to understand the feline friend who resides in your home. Even when this is not feasible, it is often possible to deduce what may have happened in his early life just by studying his reactions to individuals and events and how he deals with them. Knowing and understanding how the first few weeks can shape a cat is critical for any committed cat lover.

⚠ **It's not always easy to predict what sort of cat a non-pedigree kitten will become when he grows up.**

Crucial early weeks

During his first two to seven weeks a kitten is learning faster than at any other time in his life. It is no wonder that this sensitive socialization period when habituation also occurs, if environmental conditions are right, plays such an important role in any cat's life.

Socialization – this is the process by which cats learn about themselves and the other individuals they meet (human, feline or otherwise). This is why if they are brought up with gentle, cat-friendly dogs, cats can often form enchanting and mutually satisfying relationships.

Habituation – means learning to ignore non-threatening environmental issues, such as different people, pets and noisy household equipment, because they are encountered in a non-intimidating way.

Learning is a lifelong process

Cats continue to learn after this early period when they are relatively incautious. Everything that happens to them after they develop an adaptive fearful response to keep them safe (at about six to eight weeks of age) will teach them something, although you cannot fully make up for earlier limited or poor experiences and you just need to do the best job you can in the circumstances. If you know that your cat has missed out at this early stage, it is often unfair to try to force him into the role of family pet for which he has not been equipped.

Breed and personality

No matter who his parents were, or how his siblings turn out, your cat is special with his own personality traits. If he is a pedigree, he will also show the characteristics associated with his more limited genetic 'blueprint'. Many owners are attracted to a breed, but whatever their charming specific traits, remember that no one Siamese or Birman is going to be exactly like any other. Much depends on family characteristics and socialization, but it is often remarkable how strongly a breed's character features in pure-bred offspring or where an unsupervised mating took place between a pedigree and a moggie.

The right breed

If you want a pedigree cat, assess the breed profile before taking the plunge. Some pedigrees are bright, bold, assertive and territorial by nature. They may not fit in if you live in an area with a high feline population or a busy household where you can't dedicate enough time to ensuring that you offer enough mental and physical stimulation to keep such 'high maintenance' cats on an even keel. Neighbourhood wars can result if and when unwary owners choose enchanting and beautiful but challenging breeds, such as the Burmese or Bengal; while shy, retiring Russian Blues do not always cope well in busy homes.

🔺 **Breed and individuality combine to form the character of pure-bred cats like this handsome Bengal.**

🔺 **It is unwise to choose a pedigree cat on looks alone.**

The re-homed or rescued cat

SKILL
LEVEL

If you are adopting a cat and can meet him with the relinquishing owner in the household he has been accustomed to, this is so beneficial as it gives you a great deal of useful information to help settle him into your home. However, whether your cat comes with a reasonable knowledge of his previous history or you have to proceed from a point of complete ignorance, the principles of settling in a new pet are the same and you will need to be understanding.

● Many lovely 'rescued' cats are looking for homes with loving owners.

Settling your cat

- Always provide a safe sanctuary, somewhere quiet with everything your cat needs – some beds, scratching posts (appropriately located), toys, food, water, litter trays (as far away from each other as possible – preferably place food and water higher on a small table or chest of drawers) and at least two hiding places (more if possible). Choice gives cats a reassuring sense of control. Make them high and/or dark, perhaps an open wardrobe or airing cupboard – check for escape holes under floor boards first – or under a bed with the covers draped low, or cardboard boxes and tunnels.
- Install a commercial pheromone diffuser to create a reassuring scent environment for your new cat.

● A stress-reducing environment is essential to help your new cat settle into your home.

- Close any curtains or blinds until your cat is settled and feels confident. Then open them so that he can watch what's going on outside.
- Never rush things or try to force your cat to face new situations before he is ready – doing so is unkind and will probably backfire on you.
- Let him hide for as long as he needs to – if he eats his food, uses his litter tray (and all appears normal), occasionally moves his toys when no one is around, so be it – this is often a sign of progress.
- If he's bold enough, play with fishing rods or some thrown toys; if he's shy, sit quietly reading or chatting gently, so he gradually becomes orientated to his new family in his own time.
- If he is not used to a range of people, introduce him to the person(s) who most resembles those he was familiar with in his previous home or rescue centre.
- Unless you know he is confident with children, make sure he is really settled before they spend any time in close contact, and always use a selection of toys to give him control over the situation.
- Proceed cautiously, only gradually enlarging his physical and social worlds. Before letting him out, wipe his face and flanks with a soft cloth and use it to anoint furniture and fittings at cat height, so his new home smells reassuringly familiar. Always let him retreat to his refuge if he wants to.

The great outdoors

Once your cat's behaviour indicates that he feels settled and bold enough to venture outside, examine your outdoor space with your knowledge of feline needs to ensure it's not intimidating. Supervise his first outings, which should take place near mealtimes, so you can easily tempt him back inside if he's enjoying himself. Never leave him outside until he is well adjusted and orientated and knows his way home.

Help your cat adjust more quickly by taking things slowly and steadily.

Realistic expectations

SKILL LEVEL

Whether yours is a one-cat or a multi-cat household, you will need to develop realistic expectations based on a proper understanding of feline needs. Always be prepared to compromise, as cats have a limited ability to do so. If you don't think things through sufficiently and stick consistently to your house rules, you could be unwittingly unkind; nowhere is this more important than if you share your home with more than one cat, especially if they are individual co-habitees rather than a well-bonded feline social group

⬆ **A successful multi-cat group requires owner understanding, sensitivity and a lot of hard work.**

Establish ground rules

It is always important to understand your cat's motivation in order to avoid difficulties. Everyone must decide where he is allowed to go and stick to this. If you make changes, try to reduce the associated confusion and stress, particularly if they result in a loss of choice, prized resources or less space for your cat.

If he is normally allowed access to a spare bedroom, which is now needed for guests, make sure you close the door before they arrive and keep it closed. Install a commercial pheromone preparation and put a scratching post nearby as an outlet for his barrier frustration.

Adopt distraction strategies

To prevent your cat doing something, distract him into a different action that he finds satisfying. Keep a supply of small toys in strategic locations around your home. If he approaches 'a forbidden zone', throw a handful in the opposite direction. It is more difficult to stop

No confrontation

Try to avoid difficult and/or confrontational situations developing by getting your cat's environment and management regimes right in the first place and also by taking into account his personality and sensitivities when you or other people are interacting with him.

him performing natural behaviours, such as jumping onto furniture or chasing objects that move – provide alternatives, such as high places where he can sit and watch while his stress hormones decrease. If you are busy, give him something interesting to play with.

Observe your cat

If you are unsure how your cat feels, watch his actions and body language. Does he appear anxious? Is he indulging in stress-reducing displacement behaviours, such as over eating or grooming? Coat care is normally reserved for waking up or when a cat is relaxed, so it's a useful indicator that he is feeling aroused or unsure of himself. Always act quickly to allow him to withdraw, or just give him some more space or stop your guests and children trying to pet him.

If this is an on-going problem, you must examine your home and lifestyle from your cat's perspective to understand what is motivating this stress-related behaviour and make any changes accordingly.

🔺 **Help your cat cope with change with careful planning and sensible practical measures.**

No punishment

Never punish your cat nor think that a water pistol is a 'mild' deterrent – these stop behaviours because they are unpleasant. What may be insignificant to a bold cat may be terrifying to a timid feline. Punishment makes owners inconsistent and unpredictable, a stressor for sensitive, well-bonded pets.

59

The successful multi-cat home

If yours is a multi-cat household, then knowing what makes a feline group genuinely bond together or remain as separate individuals sharing a space is critically important. Understanding what motivates feline behaviour and the individual cats that live with you are the first steps towards getting things right, as are good observation and enlightened flexibility.

⬭ **Some cats are natural loners and will never cope with communal living.**

Selecting and introducing cats

- Sensible choice of cats is important for a multi-cat household to work in their terms. This means adopting related cats that have always lived together or carefully introducing a suitable new companion to a resident cat.
- When unrelated adult cats are brought together under one roof, they are unlikely to bond, so don't try to force the issue. This may backfire and cause distress.
- If they have the right temperaments and backgrounds and you take care to proceed slowly and cautiously with introductions, such cats may be able to live peaceably together as two separate feline social groups sharing a home. They are not one 'happy feline family' and will often be stressed if expected to share the same facilities. Eating and drinking together can be very difficult for independent cats, so watch who rushes forward and who hangs back before approaching the feeding station. Examine body language and look out for one cat sitting in a strategic location or staring at another in an attempt to distance him.

Providing the right environment

Decide who likes to hang out where in your house and garden, and provide duplicate sets of everything your cats need in each area, with hiding places and high vantage points where they can be separate in 'communal' areas, such as halls, passageways and landings. The ability to act independently, keep out of each other's way and pass by while each pretends that the other isn't there is key to keeping your cats contented.

Avoiding competition

If you have a truly bonded multi-cat group, bear in mind that as kittens mature they become more independent. And even when they retain plenty of positive affiliative behaviour (nose-to-nose greetings, allorubbing and mutual grooming), they are still likely to spend more time on their own as they age. Be sure to provide a number of beds and hiding places and split up their feeding stations; you must avoid competition at all costs.

Checking for signs of stress

Don't make the mistake of thinking that lack of obvious aggression equals feline happiness. Many cats simply aren't predisposed to such overt behaviours but they may still be stressed. Watch out for the classic warning signs, including turned backs, closed eyes, staring and body blocking. Take immediate and appropriate action to relieve the communal tension.

Defusing situations

Don't physically separate cats – it's risky and associates you with highly negative emotions that may undermine your relationship with both. Use distraction with toys to defuse a situation by increasing the distance between your cats and reducing the risk of aggression.

Play fights

These are rare in adult cats, and often indicate that all is not well. Analyze your cats' relationships and review your environment and management skills to improve things in the interest of their welfare.

⬤ **Warning signs of tension should never be overlooked if yours is a multi-cat home.**

◀ **Siblings often become more independent with age even although they remain closely bonded.**

Skill 5: Empathy

For a species that spends so much time just resting and sleeping, cats have an enviable ability to keep themselves fit and mentally engaged. Given the right conditions, they expend their daily 'time budgets' in many meaningful activities that burn up calories, produce endorphins (the body's stress busters) and keep them informed about their world. Nature has equipped our cats with a range of skills that not only enable them to thrive but which they also need to use regularly to keep themselves physically and mentally healthy.

Providing the right conditions

The key to owning a healthy, contented cat is supplying 'the right conditions', and this is where you come in. Your cat, skilled as he is, can only do so much, especially if he leads an indoor-only life – the rest is up to you as his guardian. An essential skill you should hone to perfection is providing an environment that properly reflects his needs. It will require constant adaptation throughout his life to cater for his changing abilities and sensitivities.

 Cats always like to know what is going on around them.

Exploration and energy

Exercise and appropriate mental stimulation both help to guard your cat against the negative effects of many common modern ills: lack of fitness, obesity, frustration and stress. Encouraging him to fulfil his potential and indulge all aspects of his feline behavioural repertoire brings opportunities to challenge his intellect and engage in pleasurable interactions. Along with the other skills you have developed, playing your role well in this respect should help you avoid the problematic behaviour that affects bored and under-stimulated cats.

High arousal causes problems

Accidents happen all the time but damage to your prized possessions is more likely to occur when the facilities for 'legitimate' feline activities are lacking, inadequate or so old that they long ago lost any ability to provoke interest. Cats knock things off shelves and tables when climbing, investigating, retreating or pushing light objects

Hunting behaviour

Bear in mind what your cat is learning when he plays with you and the others he shares his home with. Hunting behaviour, in particular, is a deep-seated skill that lurks inside even the most sedentary cats. Denied a harmless outlet in the form of attractive feline-friendly toys, it can be directed by an aroused and active cat towards human arms, hands and feet!

'Hunting' kittens, like this one, must never be allowed to play with human fingers and toes.

around with a paw for fun. They also check things out by chewing, and frustration-related clawing will quickly spoil your wallpaper, furniture and carpets when scratching facilities are absent or poorly located.

Aggression

This may be directed towards people or other pets. It tends to develop when cats learn to play inappropriately because their playthings are dull or unfit for purpose. They become frustrated when they have no appropriate outlets for their high arousal, although over-stimulation can be equally destructive for feline wellbeing. When cats get too 'wound up' during boisterous or repetitive games, pestered by a child or kitten, harried into a game or chased when not in the mood, everything can flip over into conflict. To avoid this, make your home and garden as interesting as possible from your cat's perspective and use his everyday routines to provide him with engaging activities, entertainment and exercise.

Happy explorations – this cat is making the most of the big outdoors.

SKILL
LEVEL

Cute kitten

You need to have realistic expectations of your cat, especially when he's a kitten. Enchanting and engaging as they are, kittens can be hard work as well as great time wasters! However, their fun and games have a serious purpose as they develop their adult physical and mental skills through exploration and play, thereby learning about their world and forming relationships. You can play your role as a responsible owner by ensuring that your kitten has these opportunities and by keeping him safe and satisfied.

Important 'don'ts'

- Don't let your kitten's activities put him at risk.
- Don't let him pester an older cat – keep him meaningfully occupied and provide the conditions for him to entertain himself.
- Don't get two kittens hoping they will keep each other absorbed – they need the same social contact and attention to their environment as a 'lonely only'.
- Don't over-stimulate or over-tire your kitten, especially if you have young children. When his energy runs out, he will need to rest.
- Don't play games with fingers or toes – it's fine now his teeth and claws are small; it won't be when they hurt!

Early learning

In their early weeks while their nervous systems, intellects and bodies are maturing fast, kittens change dramatically from day to day. Although, initially, their attention is directed mainly towards their mother and litter mates, at around seven to eight weeks, social play gives way to more object-oriented behaviour, which makes sense in a species that is known for its predatory prowess.

Socializing your kitten

Between two and seven weeks (the socialization and social referencing period), your kitten needs to have the right experiences to be groomed for his later role as a pet. This process does not stop when he leaves his mother and goes to his new home. Ensure he has the right handling and environmental experiences to keep him free from fear, sociable and well balanced as well as providing for his physical health and development. If he had a poor start in life or limited experiences before living with you, proceed sensitively and cautiously.

What your kitten wants

- Gentle handling by men, women and children of all ages and appearances.
- Sensitive and careful introductions to your home, other pets and the noise of your cat flap.
- Suitable equipment – beds, hiding places, scratching posts, food and water bowls. Litter trays must be easy to access – young kittens are naturally clean but do not

Kittens need suitable things to investigate even when they have each other for company.

have great control. If the tray is not easily accessible or too far away, they may not reach it in time.

- Safe facilities for climbing, investigating and exploring – change things around frequently to boost learning opportunities and potential fun.
- Toys – lots of light and easily manipulated objects in a range of different shapes, textures and sounds.
- An appropriate diet fed in several small meals on a daily basis – consult your vet.
- Daily health inspections will get your kitten accustomed to routine handling.

Kittens learn through a combination of observation, play and experiment.

Steep-sided litter trays can be too much for tiny kittens – adapt your facilities accordingly.

65

Adventurous adolescent

You will have endless pleasure watching your kitten's personality emerge as he grows up, but this can be disappointing if you had hoped to replicate an earlier pet or a breed standard description. So many influences make each cat what he is that no two cats will ever be the same. Between the socialization period and sexual maturity (six to eight months), as well as enjoying your kitten's daily adventures and major milestones you will be learning with him – who he is, what he wants, and what he needs you to attend to and provide. Your knowledge of his species and your vet's advice will help you get things right.

Broadening horizons

Important as his home is to him, your cat will become ever more independent as he grows. Introducing him to the outside world will open up new opportunities and exciting sights, sounds and smells, but it can be a dangerous and intimidating place, particularly if your cat is competing with others for limited space and resources.

 Apparently just resting, cats are still very aware of everything that's happening around them.

Sensible precautions

If you want your cat to use a cat flap to come and go as he pleases, don't just expect him to do so. Some cats find the noise unsettling whereas others dislike the sensation of pushing the flap, facing an environment with a restricted view and/or having to step up or down to get through it (provide a step). Get your kitten accustomed to the noise when he is having fun, playing or eating nearby. As he gets bigger, attract his interest with treats or toys. Prop it open initially and reward him for putting his head out, then for being more adventurous. Don't force the issue, as this is unkind and may set up a negative reaction from which he never recovers.

Examine your garden from your cat's viewpoint. Get down to his level and identify the open and intimidating spaces, then see what you can do about them. Move large plant pots and garden furniture near to your back door or cat flap to provide hiding places; see where you need to plant shrubs for a more natural environment; and locate potential latrine sites, ensuring they are pleasant and your cat can reach them without being seen.

Braving the unknown

Never just shut your cat outside – always plan and supervise his initial outings. Condition a 'come home' signal – rattle his dry diet container or tap a fork against his food tin. Once he starts going solo, you will have a means of attracting his attention when it's 'home time'. Begin his 'going it alone' outdoor explorations when there's plenty of daylight and near a mealtime.

Garden rules

Identify and deal with any potential hazards (see pages 88–89) and provide plenty of high places to climb to, observe from and hide and rest in. If there are lots of cats around, try to stop them coming into your garden by blocking up any holes and placing flimsy trellis along the top of fences.

◐ Cool, confident cats are happy dozing in the open.

◐ Cats enjoy going out more when they feel in control of the situation.

Well-adjusted adult

SKILL
LEVEL

It's a sad fact that many adult cats seem 'old before their time'. A combination of factors – sloth, an environment that lacks stimulation, lack of a high play drive, and owners who want an indolent companion – contribute to a cat becoming too sedentary too early. This is a shame because he is unlikely to get much out of life and an inactive lifestyle can lead to obesity with its attendant health problems. As your cat matures, your role will change but you will still be important in keeping him physically fit and mentally active.

Remember and act

- Predictability is important for cats but don't bore yours to death! Fill his toy basket with things that he enjoys, frequently changing them and adding new items.

- Cater for a variety of activities, especially related to hunting. His toys need to replicate chasing, stalking, pouncing, shaking, pawing and tearing. He can run after small balls, screwed-up newspaper, feathers on a stick or furry fishing rod toys, and pounce on them and chew them. He can rake his hind legs down cardboard tubes from inside a kitchen paper roll. Be imaginative.

🐾 **Cats usually hate going in carriers unless they have positive associations.**

🐾 **You must ensure that your cat's latrine facilities suit his individual preferences.**

If you have more than one cat, ensure they have lots of their own favourite toys when they want to let off steam by playing.

- Create positive associations with people or things (his carrier, for example) – make a parcel with greaseproof paper and a small food treat (chicken or prawns). Tie it to a fishing rod and let him 'catch and kill it'.
- Food bowls are boring! Try modelling your feeding regime on nature to encourage your cat to forage and expend more calories. Put tiny portions of wet or dry food in several places; scatter dry food over a clean patio (in good weather) or the kitchen floor; lay trails of food up corridors or staircases and encourage your cat to engage with puzzle feeders.
- The climbing frame, cat tree or activity centre your kitten had may not be big, strong or interesting enough for your adult cat. Replace any out-dated equipment, beds or litter trays with adult-sized ones.
- Litter tray hygiene often slips as time goes by. Keep your cat's indoor and outside latrine areas pleasant for him to use. Don't run out of his favourite cat litter or scrimp – abrupt changes can lead to house soiling problems as can poorly located trays, inadequate cleaning or the use of bleach or ammonia-based disinfectants. If yours is a multi-cat home, have you got enough trays (one each plus an extra) and are they located where each cat has his core area?

Games and play

- Your cat needs a range of games: playing with toys, investigating new items in the home, rushing about on his own or watching what's going on. Hang some colourful mobiles inside or out to add a sparkle to his daily round.
- He will enjoy interactive play with you. When you sit down to watch television or just be quiet, take a novel item, such as a cat tunnel, paper carrier bag (remove handles), puzzle feeder or box with a ball in it, to distract him. Have fishing rod toys to hand for entertaining him.

🐾 **A happy, healthy cat can be a most rewarding companion for an owner.**

Sensitive senior

SKILL
LEVEL

Age catches up with everyone and, sadly, your cat will be no exception. However, using all your feline knowledge and skills, and being aware of how aging affects cats physically, mentally and emotionally, will help you to get things right and make your cat more contented. After all, you do not want to inadvertently undermine his quality of life during his twilight years.

Longevity affects minds and bodies

Cats, like people, age at different rates. Some successfully remain in the first flush of youth seemingly right to the end while others are not so lucky. Elderly cats are prone to develop a range of medical conditions associated with wear and tear upon the body – heart and kidney disease, diabetes mellitus and arthritis. Even if your cat is generally fit and well, regular veterinary check ups are essential. You must be very observant so as not to miss even subtle signs that something is going awry.

Unfortunately, the feline mind is not immune to deterioration, and elderly cats often have poor memories and difficulty concentrating. They are also more prone to anxieties and inflexibility, coping less well with changes

○ **Develop your healthcare skills to help you care for your cat as he gets older.**

○ **Whatever his nature and lifestyle, use your knowledge to give your senior a happy 'retirement'.**

in their physical and social worlds. Personality change is not uncommon with sociable, tolerant cats becoming less keen on interacting with those around them or being petted, and cranky cats becoming spikier!

What your senior cat wants from you

- A cat is mature between seven and ten years, then a senior citizen until he enters the geriatric stage after fifteen. Feed your cat for his stage in life and any medical conditions that require a specialist diet. Don't take food straight from the refrigerator or leave out large quantities of dry food, which loses oils quickly, reducing its palatability. Keep portions small and tasty and make them smell more tempting by warming canned and pouched food.

- Weigh your cat regularly and note any changes in his weight, food intake, thirst and mobility. Pay attention to alterations in his activity patterns and problems with stiffness or jumping up. Pain undermines quality of life and mood, so how he feels physically will affect his emotional equilibrium. Due to good analgesics, cats need not suffer discomfort.

- To a lesser or greater degree as his years mount up, your cat may be afflicted by failing sight, poorer hearing and a reduced ability to smell and taste, but you can do a lot to help. Make only necessary changes in your home, gradually and carefully, and adapt his things, so he can easily and comfortably get to them. If he is no longer mobile or able to jump well, provide ramps or steps by moving furniture around, and low-level beds and hiding places in quiet, draught-free locations and low-sided litter trays. Lay coverings, so he does not slip on tiled or wooden floors.

- Build physical and mental exercise appropriate to his abilities into his daily routine. Use puzzle feeders but make sure he can work them and that his toys are not too heavy. Encourage him to move from time to time, even if it is only from one comfortable place to another, and engage with him in short sessions of the interactive play he has always enjoyed.

Appropriate 'use it or lose it' activities help cats to age successfully and stay healthy as long as possible.

Avoiding confusion

Confusion and disorientation can be a problem, so your elderly cat may find it helpful to have different scents, sounds and textures associated with different areas – a few drops of lavender oil on a particular blanket, for example.

Confused cats

As our feline population increasingly includes older cats so our knowledge of diseases that affect them increases. Unfortunately, like humans, they can suffer from dementia. Cognitive dysfunction syndrome (CDS) is an Alzheimer's-like condition with similar symptoms and there are currently no specific tests or a cure available, although there are helpful supportive treatments available from your vet. Hopefully, your cat will never be afflicted, but it is important to be aware of the warning signs to look out for just in case.

How can you help?

Make your cat's world small and predictable. Avoid any change and ensure everything he needs is easy to use and accessible. Discuss possible supportive treatments (special anti-oxidant enriched diets or supplements and neuroprotective and psychoactive medications) with your vet. You may need to try different products to find one that suits your cat and helps ameliorate his symptoms but your efforts will be worth it. After a lifetime of companionship, the last thing you want for him is a compromised quality of life.

Dehydration or dementia?

Dehydration is an important issue, whatever your cat's age. This can have profound physiological effects as well as causing confusion. The risk of your cat becoming dehydrated without you realizing it will increase in very hot weather, especially if he lives indoors and your home is poorly ventilated, he is fed exclusively on dry food or is suffering the effects of age. Another critical factor is how you offer him water. Placing a small bowl next to a cat's food is traditional but has little relevance to feline behaviour. Cats do not naturally drink and eat in the

🐾 Get to know your cat's 'norm' to identify any significant changes.

same places. Many dislike shallow bowls, those where their whiskers touch the sides, or plastic containers that absorb odours and taste stale.

Experiment with different water containers. Offer wide, deep bowls in a variety of designs. Try different types of water, tap and filtered, and ensure that your cat has several water stations within his territory. Inside and out, provide a choice and place some high up – on garden furniture or bedside tables, for example.

Once you establish your cat's likes and dislikes, you can monitor his behaviour to spot any changes and ensure he gets enough to drink. This will help you eliminate dehydration as a cause of any cognitive change.

Signs of dementia

Vets use the acronym DISHA for the clinical signs of CDS:

Disorientation – even in familiar locations cats are confused

Interactional changes – some cats become more 'needy' towards their owners while others do not seem to recognize familiar people and pets

Sleep problems – nighttime wakefulness often combined with restlessness and increased vocalization

House-soiling – loss of previously reliable toileting behaviour is not uncommon. It is important to rule out and treat any other possible causes before deciding a cat is suffering from dementia

Activity decrease and/or anorexia (loss of appetite) may also be seen.

Never ignore your older cat's changed reactions towards other felines as well as humans.

With their owners' help, even frail elderly cats can still enjoy a good quality of life.

Skill 6: Awareness

SKILL
LEVEL

Lack of awareness by loving owners of just how different cats and people are underlies many feline behaviour problems and undermines their relationship with their pet. Failing to provide what your cat needs to stay healthy in mind and body can put him at risk of unrecognized stress. It is very important to be aware of the emotional impact of minor events as well as major changes in your home and routines to avoid being unfair or unkind to him.

Coping with change

Often it is not the huge life-changing events that undermine feline emotional wellbeing so much as minor alterations in their environment. Thus a diligent cleaner who obliterates a cat's reassuring scent marks can cause more stress than the upheaval of moving home. The insecurity caused to a cat whose ownership 'labels' are suddenly whipped away is likely to go unnoticed, and the impact will be magnified if he is experiencing other stressful situations. The accumulation can quickly make his world start to feel out of control.

🔺 Rooms, such as modern kitchens, with hard, slippery work and floor surfaces are often noisy, intimidating spaces for cats to negotiate.

Temperament and life history

These will always play a part in how well or badly any cat copes with challenges, great or small. Some are naturally more laid back, robust and adaptable characters, while others cannot cope with even insignificant alterations in their surroundings. Getting to know your cat, accepting his traits and sensitivities and catering for them are all essential in safeguarding his emotional wellbeing.

Err on the side of caution

When you are unsure if something will upset your cat, it's best to assume it may be stressful. Take sensible measures to reduce any negative impact if you can and ensure that he has ready access to plentiful high and dark hiding places. Being able to adopt his natural coping strategies should help him to 'weather the storm'.

Awareness audit

- Get down to your cat's level and see how his world looks – understand how important it is for him to have high places to escape to and watch what's going on.
- Review everything in your home and garden from your cat's perspective.
- His world comprises: physical layout; scent and noise profiles; and social groupings (two- and four-legged). Changes can have an emotional impact, and exciting events for you may be disturbing for him.
- If something smells 'challenging' to you, it could be overwhelming for your more sensitive cat.
- Noises that you take for granted – a really loud television or crockery dropped on tiles – may be ear splitting and terrifying for your cat.
- He will feel more sensitive and less confident when he is unwell, in pain, elderly, distressed or coping with changes in his home and social group.
- Everything will have more impact if your cat has an indoor-only lifestyle and cannot get away from what is happening around him.

⚠ **If your cat shows marked or prolonged signs of stress, such as over grooming, do not delay in seeking professional help. Firstly consult your vet, who may then refer your cat to a feline behaviourist.**

What stresses cats?

SKILL
LEVEL

Being more aware of what cats generally find stressful combined with a better understanding of what they need to feel comfortable in their surroundings and what keeps them happy will make you a more perceptive and aware owner. These are essential skills for any really clued-up feline guardian.

Everyday events

Whether you are retired, living at home or go out to work, you will have a daily routine. Your cat will quickly become familiar with it and may even adapt his habits to yours. Given that cats like predictability, owners whose schedules vary little from day to day often suit their pets well. Those who live more flexible lives can inadvertently upset their feline companions, especially if they expect them to socialize when they would prefer to be exploring outdoors or napping. Be realistic in your expectations to avoid inadvertently pressurizing your cat.

This is particularly important when your unrecognized interference with a cat's ability to control his activities coincides with other stressful issues. If you arrive home

⬣ Cats that have control over their actions cope better wherever they are.

⬣ Always act thoughtfully ahead of events when you are making any changes that will affect your cat.

◁ **By not imposing her attention, this owner is actually encouraging her cat to stay with her.**

smelling of other people's cats or other challenging scents, your cat may feel threatened in his own territory without you even realizing it. Be aware of the potential emotional impact of your everyday activities and take the appropriate steps to prevent him becoming stressed.

Avoid stressing your cat

- Accept that his activity pattern may not fit yours and compromise. It is natural for cats to be active at night, dawn and dusk. If you don't want to be woken up, shut your bedroom door and provide interesting things for him to do if he does not have any outdoor access. But don't deprive him of rest during the day.
- Don't expect him to radically change his routine at weekends just because you do. Keep his management regime and social patterns as predictable and normal as you can, especially if visitors take over 'his' spare room sanctuary.
- Be sensitive with household cleaning in important areas to preserve his territorial markers, or place a well-worn scratching post there. Think twice about using 'fresh air' sprays and scented candles. Leave windows open to get rid of strong odours. Don't wash all his bedding at once: launder things in rotation, so he always has something that smells familiar.
- Close doors between your cat and noisy household activities, such as vacuuming. Never trap him in a room with a blaring television or radio.

Smells and scents

Change your clothes immediately if you smell of other cats, ask your cat-owning guests to be especially respectful of your cat, and make sure you don't pet him when you have just applied perfume. Reduce the negative effects of 'foreign' scent challenge by stowing bags, suitcases, new purchases and shopping in a closed room or cupboard or throw a blanket he has been sleeping on over them.

Major life changes

SKILL
LEVEL

We are now generally more aware of the possible negative effects that any significant changes in our home or social world may have upon our pet cats. Most caring owners try hard to minimize any adverse influences, but their expectations are often unrealistic and their cats don't always adapt quickly or well. They may take a long time to settle after an upheaval, but it is more likely to go smoothly if everyone understands the importance of taking things slowly and respecting the feline way of doing things.

Home renovations

The process of remodelling and redecorating your home will also have a significant impact on your cat, removing his familiar physical and scent landmarks and assaulting his senses with powerful intruding odours. Adopt the same approach as for moving home and don't forget to rub his scents around at cat-height before you let him explore your new layout or décor.

A changing scene is stressful

Moving house is among the top five stressors for people, and even worse for cats that neither choose to move nor understand what is going on. To an uprooted cat everything seems out of control – his security has disappeared and there are none of his familiar odours and territorial markers to help him orientate himself within his new environment. The situation is much the same when he goes to a cattery or veterinary clinic, plus he will be surrounded by animals he does not know and unfamiliar smells – all this in addition to the stress of the preceding journey. You must handle such events with patience, sensitivity and kindness and do everything you can to minimize their stressful impact.

▶ Renovation inside your home can pose physical and emotional risks for cats.

⚠ Providing high vantage points and refuges can help cats adjust to change.

⚠ Fit screens or close upstairs windows – you cannot rely on your cat's 'common sense' to keep him safe.

Changing location check list

Anticipating what needs to be done for your cat should be at the top of your list of things to action.

- If you have not accustomed your cat to his carrier, do so immediately (see pages 88–89).
- When moving home, be aware of how packing will upset him. Keep one area as normal as possible as his sanctuary, and make sure his favourite bed, hiding place and so on are not disturbed; leave plenty of his reassuring scents around and use a pheromone diffuser to help him cope with the upheaval.
- Plan his transportation carefully. So he is not subjected to unnecessary stress, you could send him to a friend, relative or cattery for a few days.
- Once in your new home, sort out his sanctuary as a matter of priority (see pages 56–57). Include familiar furniture that will smell 'of home'.
- Allow your cat plenty of time to settle before enlarging his world, especially if he is timid by nature or finding adjustment difficult.

Feeling at home

If your cat spends time in a cattery or veterinary clinic, take a cardboard box he has 'made his own' as well as his bed with you to give him a comforting refuge to hide in until he feels sufficiently confident to explore. Include some clean, used blankets or towels that smell of him. And give his temporary carers some of your things impregnated with your scent to keep your bonds intact.

New additions, lost companions

Changing characters on his scene will be an important feature in your cat's life. Although they may be exciting for you, for him they are simply challenging. Therefore careful consideration of what he needs is essential to prevent a happy event being soured or a sad one made even worse. Awareness is one of your most potent weapons when it comes to protecting your cat and helping him adapt to changed circumstances. Patience is another – never underestimate the time it may take for him to adjust to newcomers.

Getting more pets

A common mistake grieving owners make if they lose a cat or dog, particularly if their remaining pet appears distressed, is to rush into getting a 'replacement', but each pet is valued for himself and can never be replicated – nor can the relationship between a cat and his departed 'friend'. If your cat loses his companion, he will be affected and may seem sad and lost, but foisting a new pet upon him is unlikely to make things better. Keeping everything as it was, comforting him with familiar routines, food treats, play or quiet companionship is the way to help. Only gradually remove familiar smelling items and change things to reduce the stress for the cat that remains at home.

New arrivals

A new partner, baby, lodger or pet will profoundly change your cat's world. As far as possible, you should keep his own items and management routines as they were – predictability will help him cope. If you must, gradually alter them ahead of any changes in schedules or in his environment to reduce their negative impact.

Introduce baby equipment sensitively in such a way that your cat is not overwhelmed by new sights, sounds and smells all at once. Play recordings of babies crying when he is relaxed, playing or eating, so he becomes accustomed positively to the imminent new arrival.

With new people, introduce some clothing they have worn next to their skin to make initial introductions low

⬥ **Always be sure to make introductions positive by associating new people with good things.**

⬤ When a cat loses a valued companion, he misses the individual rather than just 'a dog'.

key. Combine them with something your cat thinks is good. Never allow anyone to pressurize him with the aim of 'making friends'. Make them aware of feline needs and his temperament, supply them with some fishing rods, and get them to take over feeding times to create a good impression and break down barriers.

Deaths and departures

On sad occasions, the impact on pets is not always uppermost in anyone's mind, but this can be significant, especially if a cat was very close to someone. An elderly or unwell person was often a quiet and predictable presence, whose company is much valued by a resident feline. Leave some of their possessions and bedding to soften the blow for a well-bonded cat.

⬤ Whether it's an old acquaintance or a new friend, keep interactions short and sweet.

Recognizing signs of stress

Your knowledge of feline behavioural issues and awareness of your cat's history and personality will help you, if despite all your careful efforts, his resilience is undermined. Owners do sometimes slip up in believing that their cats always overtly express signs of their stress or inability to cope with what is happening to them – they do not. However, learn to recognize the subtle, telltale indicators of feline stress in order to help your cat.

Professional input

At no time will your understanding be more critical than if your cat develops a behaviour problem that could be motivated by frustration, stress or emotional conflict. Distressed cats may become aggressive or start marking with urine, or even faeces, inside the home. They can also develop severe over-grooming and compulsive (repetitive) behaviours, such as fabric or electric cable chewing, or even self-mutilating conditions.

Careful observation is essential

When assessing how your cat feels watch what he does. Frequently the signs of arousal and emotional discomfort are quite obvious, but even caring and observant owners can easily overlook them. Note any changes in his normal behaviour patterns. Is he more withdrawn than usual? Does he hide away for long periods when he would normally be interested in what is going on? Does he spend his days with neighbours rather than with you? He may be choosing to spend time in a quiet household because there is too much going on at home and he finds it hard to cope with all the activity and bustle.

🔺 **Quickly seek professional help if your cat shows signs of distress, such as over-grooming.**

Changes are often significant

Displacement activity is a good way to ascertain how your cat is feeling. If he grooms more than normal, particularly in certain situations or in the presence of some people or pets and not others, he is not entirely comfortable with them. Analyze all the evidence to understand who and/or what is unnerving him and introduce as many feline-friendly measures as you can to help reduce his arousal (see pages 44–45 and 48–51).

Marking behaviours are also informative when it comes to assessing feline stress levels, especially when associated with more intense sniffing of people's clothes or objects, fixtures, fittings and possessions. Get used to what your cat usually does and note any variations in his patterns. Increased bunting and rubbing around his environment or on those he shares his home with will tell you that he is feeling insecure. Work out why and what you can do about it. Never assume that things will get better of their own accord – he needs your help.

⬤ **Physical barriers are needed to protect your cat from potential hazards.**

Signs of stress

- If your cat shows visible stress, never punish him; he needs understanding.
- Don't fuss and pet him to make him feel better – 'keep calm and carry on' as normal. He will find that more reassuring.
- Consult your vet to check a medical condition is not underlying the problem.
- Carefully evaluate your home environment and management routines to see what could improve.
- Consult a qualified feline behaviourist if your common sense 'first aid measures' don't quickly resolve the problem.

⬤ **You should carefully supervise all interactions between unfamiliar pets to avoid potential problems.**

SKILL
LEVEL

Skill 7: Responsibility

Cat owning is a joy but it is also an enormous responsibility and a lifelong commitment. Your cat is a valued member of your household with individual needs based on his species' behavioural repertoire, his inherited and personal characteristics and everything he has learned throughout his life. That makes you responsible for his physical, mental and emotional welfare

Considering your cat's needs

Like any committed relationship, the more you put into it, the more likely you are to enjoy the benefits of owning your cat. Quiet companionship, entertaining fun and the pleasure that comes from nurturing a charming animal can be immensely rewarding for responsible owners.

⚠ Many common flowers and indoor plants are toxic to cats, including tulips. Don't put your cat at risk.

▷ Owners of cats with outdoor access should still provide a safe and interesting environment.

You will need all the skills you can acquire to help you understand your cat's ways and provide the sort of world he needs to keep him physically fit, mentally active, well balanced and happy. You also have to combine realistic expectations and practical measures with sensitivity, good observational powers and flexibility to ensure that you get things right. And learning from the mistakes that other owners commonly make is another valuable item in your 'good ownership toolkit'.

Problem behaviour clues

A common influence in many feline behaviour problems is the failure of owners to understand how their changed circumstances affect their cats. They frequently assume that cats will cope with anything life throws at them, but they can't, so you need to carefully consider the impact upon your cat of anything you propose to do. Whether you are restyling your home or planning to make some major changes in your life plan or relationships, you must take steps to protect your cat from the consequences he may suffer. Sometimes that means putting aside your own desires in the interests of his wellbeing.

There are so many different elements to responsible ownership: financial issues cannot be ducked and adequate provision must be made for veterinary bills as well as everyday equipment. Over the years these mount up and there are some things that cats can't do without, such as activity centres, hiding places, toys, beds and litter trays, especially if they are indoor-only cats. If you try to cut corners with such items, your cat will invariably pay the price. Compromise is not always easy but it is an essential aspect of being a responsible owner.

◗ **Starting regular healthcare checks early in kittenhood will make them less scary for your cat later on when he is an adult.**

After you've gone

None of us wants to contemplate our own mortality but you need to make provision for your cat in the event that something happens to you. You have a duty to repay his committed companionship by ensuring that he is properly taken care of if you are no longer around to do so. This is, after all, the last loving and responsible act you can carry out for a cat you have shared your life with.

SKILL
LEVEL

New companions

Nowhere do owners make more mistakes than when it comes to acquiring another pet – although it may be a blessing, it could be a disaster for your cat. It is difficult to refuse a needy, deserving cat a good home or to put off getting a dog when your heart is set upon it. However, your responsibility is to give priority to the emotional needs and welfare of your existing cat.

⚠ Cats sometimes develop surprising relationships.

The multi-cat home

This is one of the most challenging feline issues of our time. Cat lovers love living with cats but not all cats can live happily with others. Before expanding your feline group or getting a new cat when you lose one, think carefully about your remaining cat. Refer to your knowledge of feline behaviour, ask yourself some basic questions and answer them honestly to help you make a responsible decision. If you go ahead, make a good choice of companion and conduct a patient and sensitive introduction; 'letting them get on with it' or just hoping for the best is simply irresponsible.

⚠ Even well-bonded cats can sometimes find their companions too close for comfort.

⚠️ **Having enough space for quiet retreat and withdrawal helps cats to avoid conflict.**

Introducing a new cat

Settle him first (see pages 56–57). Once he is calm and has found his feet, you can start to swap scents between your two feline groups. Follow these simple guidelines.

- Proceed slowly, always supervise, carefully observe the cats' reactions and never pressurize.
- Use a separate cloth to wipe each cat. Offer this to the other when you feed or play with him.
- When this is accepted without any negative response, mix the cloths together in a plastic box or bag and repeat the procedure,
- When your cats seem to accept this calmly, let the newcomer explore your home when your resident cat is outside or sleeping quietly (shut the door).
- Eventually when your cats are ready to meet, leave nothing to chance.
- Provide lots of hiding places and other facilities to avoid competition for resources, and place toys all over the house to distract them from each other.
- Take things slowly and always be realistic and fair.
- Always be prepared to slow down or go back a step or two if necessary. Don't rush things.
- If things are not working out don't persist – contact a qualified feline behaviourist.

Longer term, remember the importance of maintaining separate core areas with everything the cats need based on the places where they each like to 'hang out'.

Checklist

Before getting another pet, consider the following:

- Do you have enough space, money, time, resources and energy?
- Are there already too many cats in your area?
- Is your preferred cat or dog likely to get on with your resident cat?

Is your cat:
- Accustomed to living with another cat or dog? If so, is it the individual pet he is missing?
- What sort of start did he have in life; does he have the right temperament to cope with another pet?
- Is he territorial?
- Does he like having all your attention?
- Is he of an age when he can adapt to changed circumstances?
- What will happen if things don't work out well?

SKILL LEVEL

Keeping your cat safe

You must do everything you can to ensure your cat's physical safety. Be aware of what may be potentially harmful and be constantly vigilant to deal with any changes in your home or garden that could put your cat at risk. Complacency has no place in the responsible cat owner's repertoire.

Carrier comfort

- Keep the carrier out and condition positive associations with it.
- Take the front and/or top off, put a familiar smelling blanket inside so that it becomes a comfy bed, and place it where your cat likes sitting.
- Feed him nearby, play in the vicinity and hide some tasty treats inside.
- Repeat this procedure in several places.
- When your cat views it positively, put the carrier back together and repeat the process until he is happy and relaxed spending time inside.
- Spray inside the carrier with a commercial pheromone spray before outings, and cover with a cloth to make the experience less stressful.

Safety first checklist

- Check your environment for potential hazards – ensure your cat cannot crawl into spaces he cannot withdraw from – under floorboards or inside cupboards. Fit screens to high windows, cover garden ponds and check for toxic plants (indoors and outside).
- Store everything, including medications, cleaning materials, rubbish and food waste, safely.
- Use non-toxic cleaning agents or chemicals.
- Restrict your cat's access to affected areas.

Identification

Get your cat microchipped – your vet will implant one painlessly under his skin. Its unique number is registered with a central computer, so you can be reunited if he

🔺 **Microchips often reunite owners with cats that have got lost or gone missing.**

If your cat wears a collar choose it for safety, not for its aesthetic appeal.

wanders. Don't forget to change your contact details if you move home. Consider an additional means of instant recognition, such as a collar with an identification disc. However, only use a design with a quick release mechanism in case your cat gets caught up by it.

Travel safety

Provide your cat with a well-designed, well-maintained carrier. Always clean it after use to remove any alarm pheromones deposited when travelling. Before any journeys put in a new blanket, preferably one he 'has made his own' when he was relaxed, as this will be comforting. Don't keep the carrier out of sight until you need it or it will quickly acquire 'predictive value' and act as a negative signal when you get it out.

Leaving your cat alone

Whether you leave your cat alone during the day or for longer holiday periods, you must consider his needs, especially if he lives solely indoors. It is crucial never to take chances with his safety, under-estimate the negative effects of boredom, or neglect your role as a responsible owner in providing interesting and stimulating things for him to do while he is alone.

○ **Holiday care – your cat's individual needs, temperament and history should determine your arrangements.**

Daily fun and games

Whatever his age, your cat needs physical exercise and mental stimulation and the right conditions to indulge these needs. Providing a safe and interesting world need not be expensive. Just use your imagination and some recycled cardboard boxes and tubes and old furniture. Anything you buy must be stable and safe and not treated with toxic preparations or paints.

Holidays

When you're going away for longer, do make proper arrangements for your cat. Leaving food and water for a couple of days is not responsible – cats may suddenly become ill or have an accident. Take your cat with you, have someone care for him at home or send him to a cattery. Consider all the practical and emotional issues before deciding. If your cat is timid, territorial or has recently experienced upheaval, leaving his familiar home territory may be more distressing than staying there. If he is people-orientated or needs medication, unless someone who knows him can move in to keep him company, a cattery may be your best option.

Responsible preparations

Take your cat on holiday with you only if he will be safe and has everything he needs for his physical, mental and emotional well-being. If you opt for homecare, only consider responsible, experienced cat sitters to live in or visit daily. A professional is likely to bring the scent of other felines into your cat's home, which is stressful. The same holds for cat-owning neighbours, whose visits may encourage incursions by their cats, another stressor for yours. Often the best cat sitters are knowledgeable people without cats themselves. Starved of feline affection, they relish the chance to spend time caring for a cat that may feel lonely. When choosing a cattery, ask for recommendations from other owners and your veterinary clinic. Always visit beforehand to ensure that it is suitable and the staff are responsible and caring.

Holiday check list

- Take his own things if your cat goes away with you or stays at a cattery.
- Leave clear instructions with carers about his likes and dislikes, normal routines, details of your vet and medical history.
- Don't wash his bedding or yours before you leave home – familiar odours will be reassuring.
- Leave items of clothing you have worn, sweaters or T-shirts, or towels you put in your bed in zip-up plastic bags to 'top-up' your scent bonds.

⚠ **Allow your cat to quietly readjust to your presence in his own time when you have been away on holiday.**

Visiting the vet

SKILL LEVEL

Register your cat with a cat-friendly veterinary clinic, find out about emergency out of hours arrangements and what you need to do for his routine medical care. Veterinary visits are often distressing experiences for cats, so you have an important part to play in making them as stress-free as you can for your pet.

⚫ **Vet visits will be less intimidating for your feline in cat-friendly clinics.**

Careful preparations

Be aware that not all veterinary establishments are geared up to handle cats as well as they deserve in what is inevitably an intimidating environment.

- Ask for recommendations from other cat lovers.
- Check waiting rooms for feline-friendly features, commercial pheromone diffusers or shelves for cats, so they don't have to wait on the floor.
- Enquire about staff qualifications to find out if they undertake specific courses related to cat care or have won any cat-friendly awards.
- Once registered, keep all your cat's vaccination certificates and other paperwork safe in a handy folder in case of an emergency.
- Note in your diary when regular treatments are due.
- When preparing for an appointment, keep your cat inside the house ahead of time, so you don't end up loading him into his carrier in a rush.
- Transport him carefully with his carrier covered.

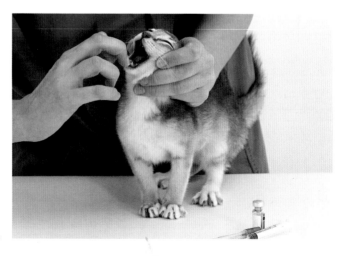

▶ **Carefully transporting your cat will help reduce veterinary-induced stresses.**

⚫ Cats deserve carefully conducted stress-free partings. Be sensitive but clear-headed when making decisions.

Multi-cat homes

Special care is needed if your cat lives with other cats. It is not uncommon if one cat returns from a veterinary visit smelling clinical for other cats to react badly. Evidently he no longer seems familiar and probably evokes bad memories. It may help to reduce this risk if you wipe 'the patient' with a soft cloth before leaving home. Store it in a zip-up plastic bag and wipe him again to familiarize his scent before you return.

Once back at base, supervise initial meetings between your cats to observe any warning signs. And have the means – fishing rods or toys you can scatter around – to distract their attention away from each other. If things go wrong, immediately separate them, let them settle down for several hours at least and try introducing them again in a very low-key way. Do not persist if tensions continue to run high – seek professional help.

If one cat is away for some time, for example in hospital, don't just expect to start again where you left off in relation to your feline group. You may need to conduct a careful re-introduction programme as you would if bringing a new cat into your home.

Letting go

Veterinary science has made enormous strides in recent decades, so when caring for our cats it is remarkable what can be achieved. However, at some point every owner has to face the sad reality that their cat's life is drawing to a close. When the end comes, quality of life is the most important issue and your vet should support and advise you if you are faced with making difficult decisions about euthanasia. It is wise to think ahead about how this may affect you both and decide whether or not you will stay with your cat and what to do afterwards. The same applies to whether you bury him at home or get the clinic to arrange cremation or burial at a pet cemetery. Making these decisions when you are upset could prolong things for your cat, and risk you feeling later on that you did the wrong thing. Discuss everything in advance with your vet, so you are prepared, then together you can make the whole process as stress-free as possible for your beloved cat.

Bringing it all together

Throughout their long association with humans over many centuries, cats have attracted and fascinated countless numbers of people. Many of us find great joy in sharing our homes with them but even so there have been times when their behaviour has caused controversy. Sometimes it still does, within a household, neighbourhood or wider society. It can certainly be perplexing, even to those who believe our lives are enriched by having feline friends. One word – mysterious – is often attached to the behaviour of the cats that live alongside us. However, as you have seen, this need not be the case.

⬛ Sensitively applying your esential skills to your cat's care will provide benefits for you both.

No longer a mystery

Once you have realized that we have had surprisingly little influence on our cats and examined the behaviours their forebears developed to survive and multiply so much becomes clear. There is still a lot to learn but science is leading us towards a better understanding of feline senses and needs and showing how different they are from ours. As a consequence, your cat's view of his world and those around him, whatever their species, differs from your outlook. He may be negatively affected by everyday items around your home and garden or find stressful things and activities that you take for granted.

Understanding the importance of genetic inheritance, personality and early experiences will help you work out why your cat takes some things in his stride but is unsettled or 'spooked' by something he encounters frequently or on an occasional basis. Even if you have no idea where he came from or what his immediate ancestors were like, you can isolate the elements of your cat's individual behaviour from his inherited species' traits using your newly found knowledge.

Cat-friendly environment

You know now how crucially important a 'cat-friendly' environment is to your cat and what he needs you to provide for him to make your home, outdoor space and management regimes just right from his perspective. You are aware of how minor changes in the environment or

his daily routines can affect him emotionally, and how care is needed when, for example, cleaning your home, making alterations in your décor or undertaking major life changes to avoid inadvertently causing him the sort of distress that can lead to problematic behaviours, or in a worse case scenario undermine your relationship.

Understanding his behaviour and what he needs from his world will help you provide exactly what your cat wants.

Now you know what your cat wants

Combining greater comprehension and understanding of all these important issues with empathy and awareness will enable you to accurately identify your cat's motivation, whatever he is doing. You will be able to anticipate, plan and provide well for him every day and when it comes to the large and small challenges of life. All your skills and affection plus your desire to act responsibly will come together to help you ensure he is as healthy, well adjusted and happy as he could possibly be. In short, you now have a greater chance than ever of giving him just 'what your cat wants' out of life. And that surely has to be one of the greatest and most rewarding joys any cat owner can experience.

Index